Process Development Symposium 2014

Solving Today's Global Challenges

Philadelphia, Pennsylvania, USA
10 – 12 June 2014

ISBN: 978-1-5108-1759-3

Printed from e-media with permission by:

Curran Associates, Inc.
57 Morehouse Lane
Red Hook, NY 12571

Some format issues inherent in the e-media version may also appear in this print version.

Printed by Curran Associates, Inc. (2015)

For permission requests, please contact AIChE
at the address below.

AIChE
120 Wall Street, FL 23
New York, NY 10005-4020

Phone: (800) 242-4363
Fax: (203) 775-5177

www.aiche.org

Additional copies of this publication are available from:

Curran Associates, Inc.
57 Morehouse Lane
Red Hook, NY 12571 USA
Phone: 845-758-0400
Fax: 845-758-2634
Email: curran@proceedings.com
Web: www.proceedings.com

TABLE OF CONTENTS

SUSTAINABLE PROCESSING

EMERGING TECHNOLOGIES

BEYOND TECHNICAL KNOW-HOW – MASTERING SOFT SKILLS

Developing Processes in a Changing World — Future Trends that Will Change The Way We Work.

Henry T (Hank) Kohlbrand
Chemical Industry Consultant
2010 AIChE President
Global R&D Director, Engineering & Process Sciences (Ret)
The Dow Chemical Company

Process Development and AIChE

- A journey that began in the early 1980's

Introduction

- Examining key trends and drivers helps us to prepare ourselves for the future
- Change is all around us, but it has always been a part of chemical engineering careers
- Opportunities often appear at the interfaces of science and where dislocations appear
- With globalization has come increased complexity for decision making as multiple objective functions come into play

Slide 3

Drivers

- Globalizing to Globalization
  - Economic emergence of new economies
  - Movement of assets to low-cost regions
  - Political stability becomes more important
- Changing business models – companies are changing to assure future success
- Standard of living increasing globally
- Increasing interest in a sustainable future
- Energy is at the center of many discussions today
- Water and food will be at the center of many discussions tomorrow
- Industry expanding into geographies where energy and water are scarce

Slide 4

A Look Back in Time – Disruption vs Opportunity

- 1970 EPA established/Clean Air Act
- 1973 Oil Embargo changes the cost structure of petrochemicals and fuels
- 1974 Nypro Flixborough explosion sends a safety wake-up call through the industry
- 1984 Bhopal catalyzes cooperation and raises the bar for performance expectations
- 1990 Impact of globalization accelerates
- 2001 9/11 changes plant security
- 2001 U.S. Chemical imports exceeded exports
- 2012 Shale oil and gas change the landscape

Design Heuristics Impacted

Emerging Markets Have Emerged

Courtesy: Bill Banholzer

Ethylene Capacity Expansions-2010 view

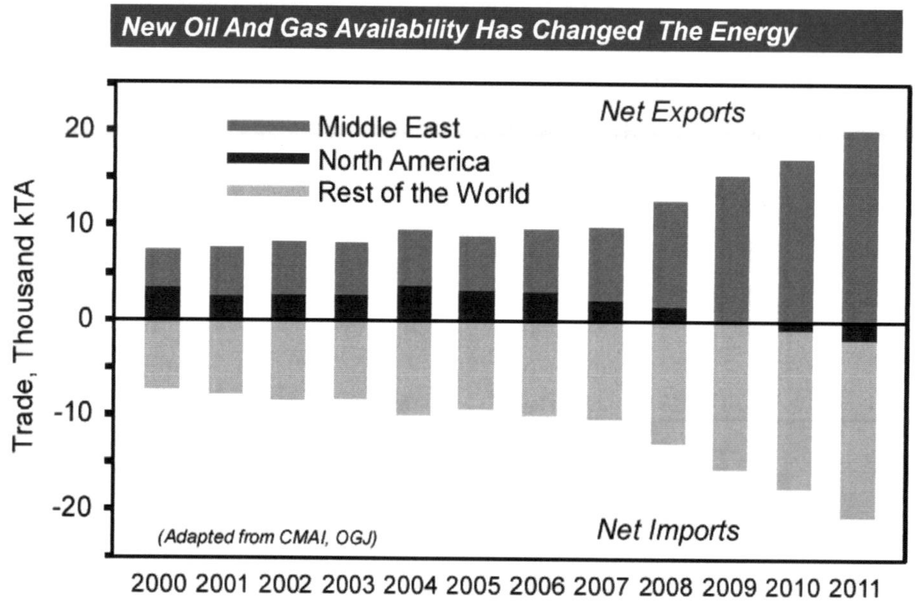

Slide 7

Challenges for the next generation of change

Energy **Climate Change** **Water** **Health & Nutrition** **Transportation & Infrastructure**

- Sustainable solutions to these challenges will provide many opportunities for current and future chemical engineers
- An eye on product as well as process will provide additional opportunities
- Sustainability will provide a strong need for innovation

Slide 8

Today's World Becoming the "Crowded Planet"

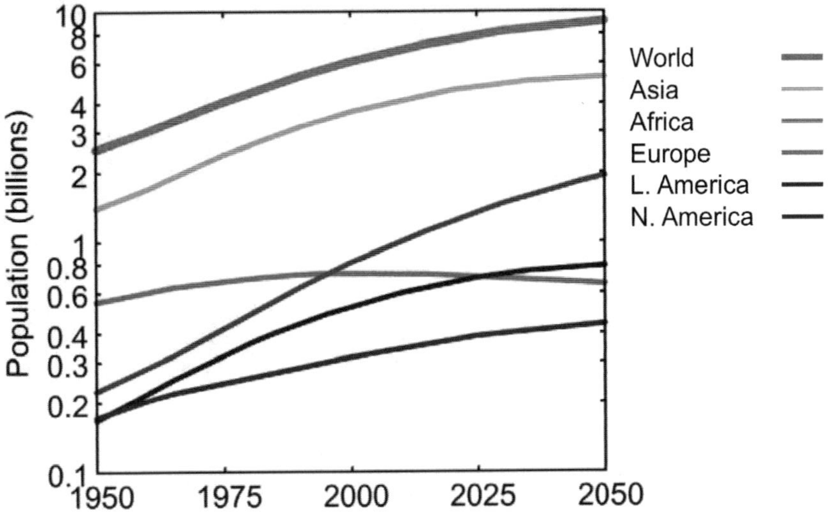

Courtesy: John Sofranko

The "Crowded Planet" Creates Today's Grand Challenges

- 2.7 billion people live on less than \$4/day and cannot access electricity, clean water or sanitation
- Growing demand for energy / the ways energy is used will change
- Extensive environmental threats / issue of climate change
- Inequalities of income and power (both within and between countries)
- Rapid technology diffusion – challenges of Intellectual Property protection globally

Energy, Environment, Water, & Healthcare

Courtesy: Dale Keairns

Global Electricity Usage

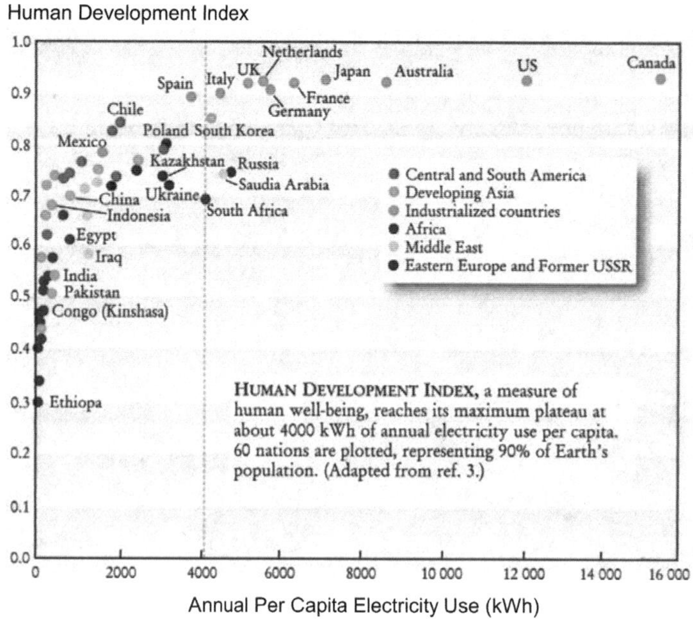

Energy demand could triple by 2050

Courtesy:

June Wispelwey

Source: United Nations

Slide 11

Status of Energy Technologies

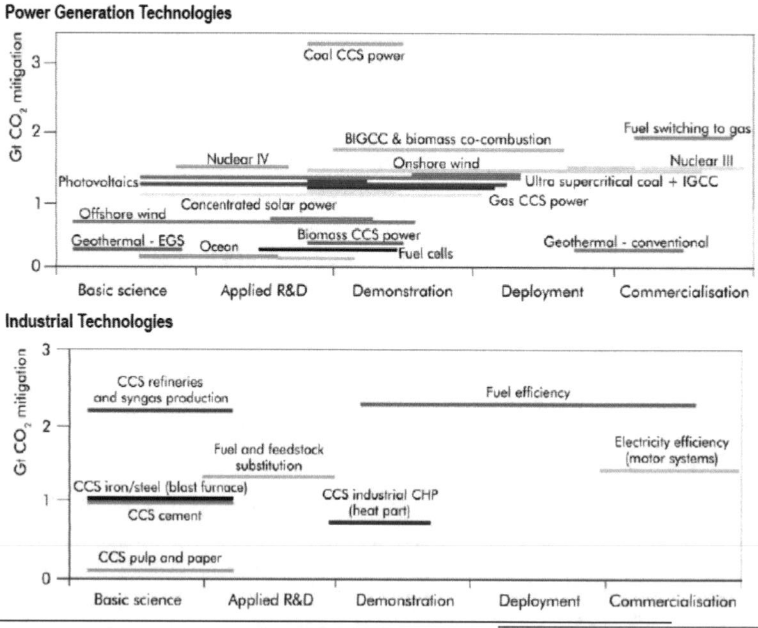

Source: International Energy Agency 2008

Slide 12

Water – Opportunities Abound

- Leaky pipes result in loss of 7 B gal drinking water/day in U.S. (Science 17 Feb 12 p 777)
- Infrastructure and prediction of failure rates
- 1.4 B live in areas where water consumption > renewal
- U.N. – 1.8 B will live with water scarcity & 2/3 globally will live under water stressed conditions by 2025
- Over past 100 years water consumption grown at twice rate of population
- 70% use agricultural, 22% industrial, 8% drinking
- China' s annual water deficit is roughly 40 billion m3 in normal years, and half of its cities are facing some degree of water shortage.
- 85% of India' s water is used for agriculture (engaged in by 60% of population

Slide 13

Future

- Global water requirements will grow from 4,500 billion m3/yr today to nearly 7,000 billion m3/yr in 2030 — a 50% increase in only 20 years
- By 2030, available water supplies might satisfy only 60% of demand.
- A shortage of water can inhibit energy production; may result in more energy being used to pump water
- Purification of marginal quality water requires a lot more energy to purify and make available for use
- As industry grows in arid areas in developing countries more creative solutions will be required
- This issue will clearly impact the future of the processes we develop today.

14

Water –Energy Interdependence

The Water-Energy Nexus

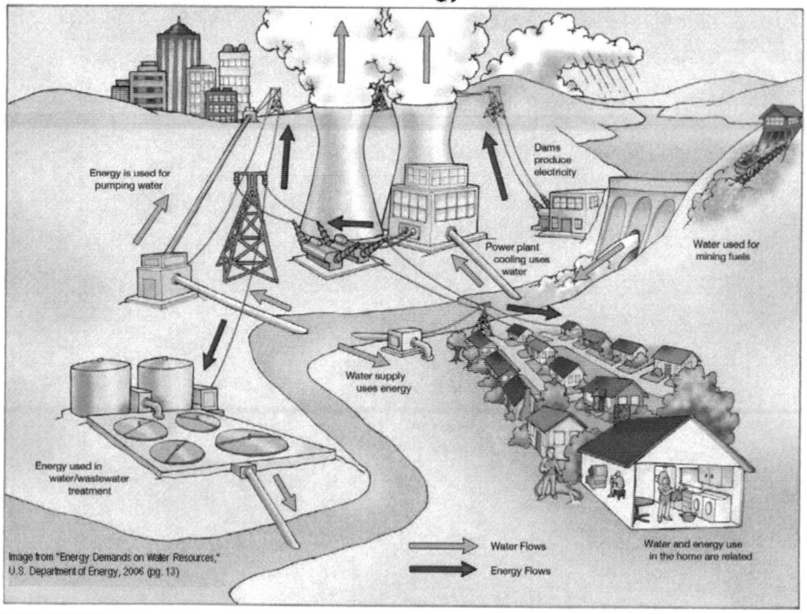

15

Water for energy production

Water consumption coefficients for energy technologies

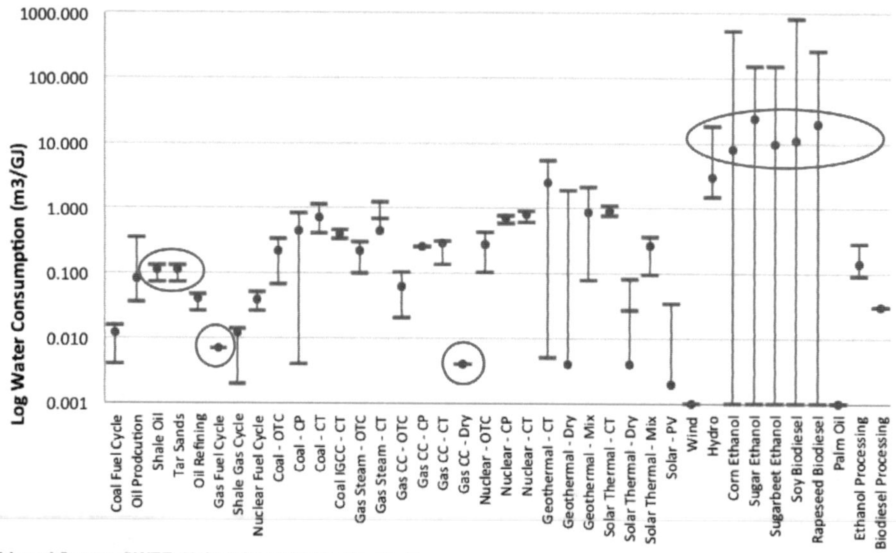

Dr. Edward Spang, CWEE, University of California, Davis
A thirst for power: A global analysis of water consumption for energy production, 2012

Slide 16

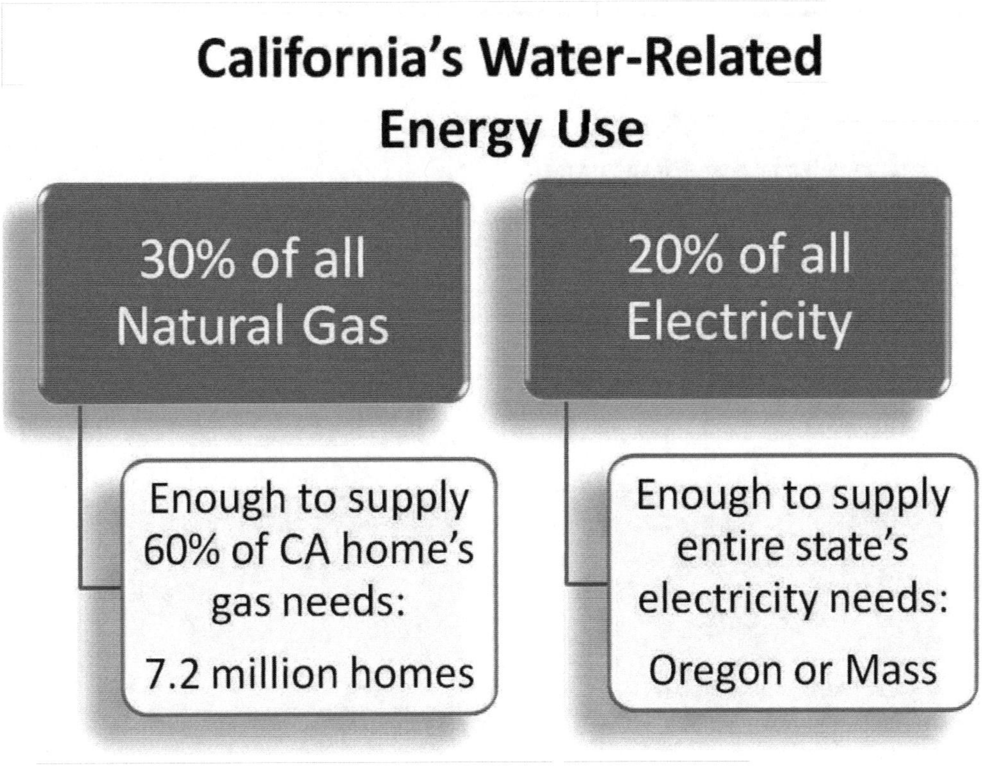
California's Water-Related Energy Use
30% of all Natural Gas
Enough to supply 60% of CA home's gas needs:
7.2 million homes
20% of all Electricity
Enough to supply entire state's electricity needs:
Oregon or Mass

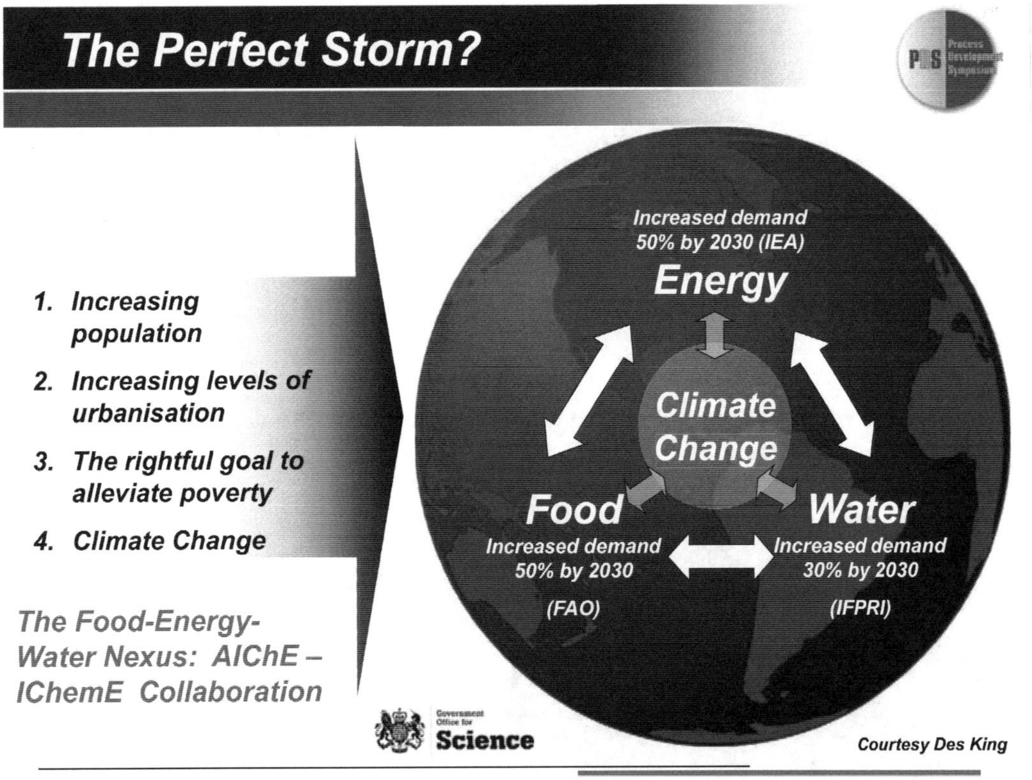
The Perfect Storm?
1. Increasing population
2. Increasing levels of urbanisation
3. The rightful goal to alleviate poverty
4. Climate Change
The Food-Energy-Water Nexus: AIChE – IChemE Collaboration
Increased demand 50% by 2030 (IEA)
Energy
Climate Change
Food
Increased demand 50% by 2030
(FAO)
Water
Increased demand 30% by 2030
(IFPRI)
Government Office for Science
Courtesy Des King

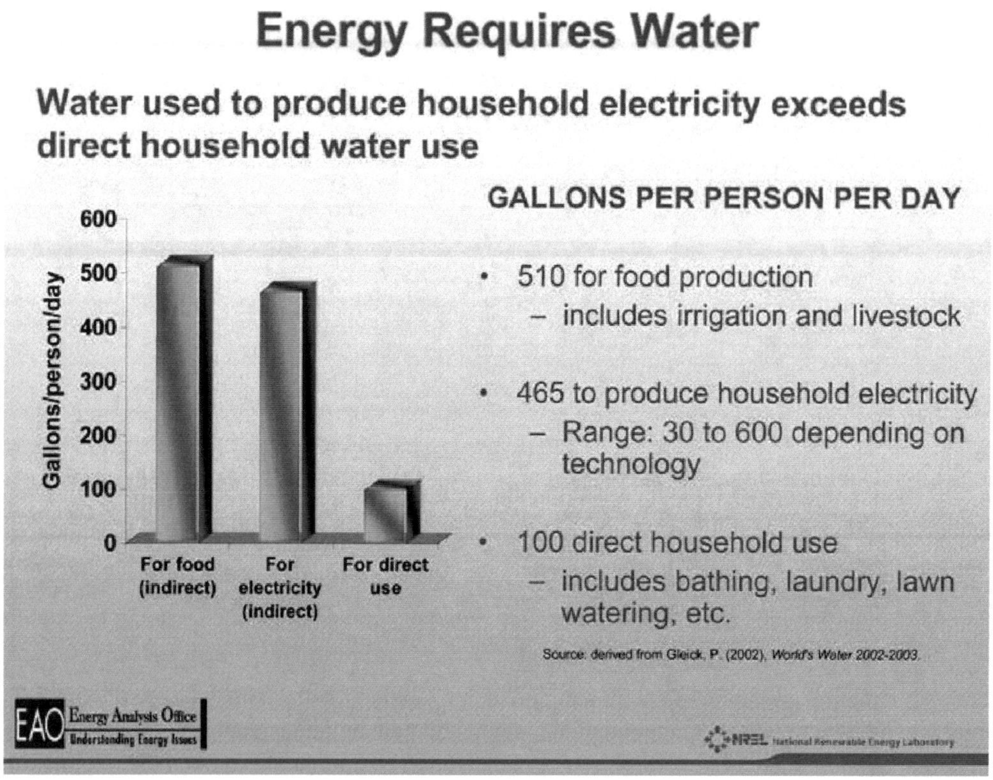

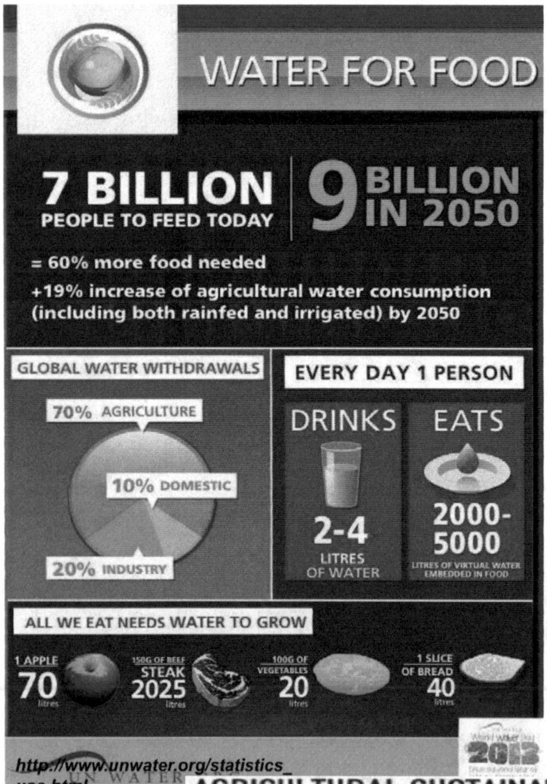

http://www.unwater.org/statistics_use.html

ENERGY FOR FOOD

Food production and supply chain is responsible for around 30% of total global energy demand

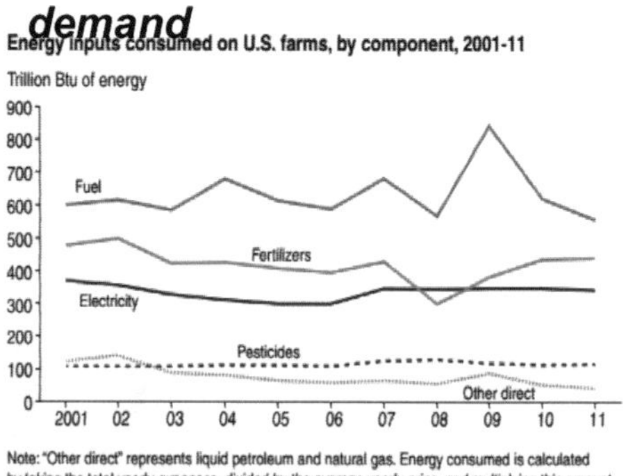

Note: "Other direct" represents liquid petroleum and natural gas. Energy consumed is calculated by taking the total yearly expenses, divided by the average yearly price, and multiplying this amount by the energy conversion ratio. Btu = British thermal units.

Source: Miranowski (2005) and USDA, Economic Research Service calculations.

Also must take into account transport, storage, processing and other energy costs for food.

Courtesy Kate Scow

AGRICULTURAL SUSTAINABILITY INSTITUTE AT UC DAVIS

Now What?

- Design processes understanding key sensitivities and connections to the megatrends
- Increase in population + better living standards creates both opportunity and liability
- We build processes for 30+ years in some cases – look over the plant lifecycle
  - Learn from the power industry
- Current pricing for water and carbon don't always reflect future market conditions
- Look for potential dislocations (ex natural gas)
- Renewable fuels and raw materials are impacted by water cost/availability and food competition
- Human needs will trump industry needs

Slide 22

Tools

- Chemical Engineering will play a key role in creating a sustainable future with foundation of good science (ex thermodynamics)
- Innovation and optimization using multiple objective functions with direct our focus
- We need to recognize interdependencies and take a systems approach
- Life cycle analysis
- Sensitivity analysis
- Process Technology

Solutions – mulitple objective functions

Process Technology of the Future

- Process design and synthesis
- New unit operations/process Intensification
- Separation network optimization
- Alternative raw materials
- Sustainable chemistry
- Breakthrough mentality (set stretch goals for important elements)
- We need different approaches and new assumptions to address these needs
- Needs to be applicable across geographies

Slide 25

New unit operations/process intensification

- Challenge the status quo
- Rearrange order of operations
- Combine reactors and separation in creative ways (reactive distillation membrane reactors, etc.)
- Hybrid separations
- Microchannel systems (large scale challenges)
- High gravity reactors
- Use advanced control for unsteady-state operation
- Objective is smaller, less expensive, more efficient systems

Slide 26

But … It's not only about process anymore

- Future grand challenges include connecting how we make things to how they perform
- Fundamental understanding is needed in key product areas
- Product performance often eludes precise measurement
- As engineers and scientists we need to focus more on product development and performance understanding – sustainability requires evaluating function and not just form
- Consider the spectrum commodities → performance → highly specialized

Slide 27

Rotating Packed Bed Reactor

- No Mechanical Issues!
- Very easy to startup and shutdown
- Yield = 94-96% (80% conventional)
- RPBs < 40 x smaller
- 1/3 reduction in waste water & chlorinated byproducts
- RPB technology has significantly improved process performance compared to conventional

Wrap up and summary

- The future is on its way
- Megatrends will shape many of our technology choices and business opportunities
- Develop processes and understand sensitivities
- Use innovation and new heuristics to develop and design successful processes

Thank you & Questions

Dynamics and Control of Benzene Hydrogenation via Reactive Distillation

2014 Process Development Symposium

Vishal Mahindrakar
Juergen Hahn

Department of Chemical & Biological Engineering
Department of Biomedical Engineering

Rensselaer Polytechnic Institute
Troy, NY

Environmental Regulation of Benzene Content of Gasoline

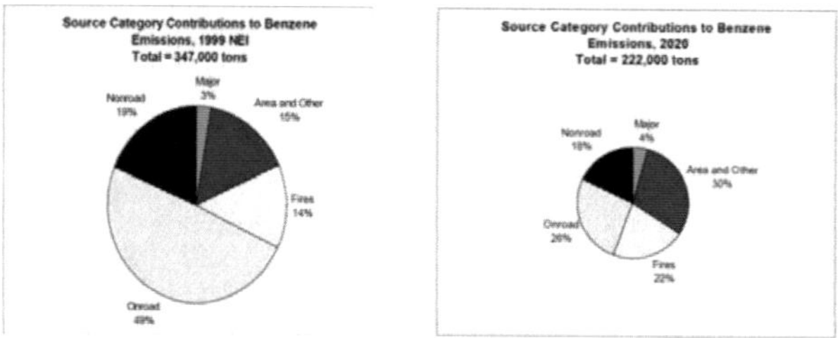

Benzene contribution of all source sectors ("SBAR Panel Report on Mobile Source Air Toxics", EPA420-R-05-901, November 2005)

Fuel Program

EPA is requiring that, beginning in 2011, refiners must meet an annual average gasoline benzene content standard of 0.62 percent by volume (vol%) on all their gasoline, both reformulated and conventional, nationwide. The national benzene content of gasoline today is about 1.0 vol%. (Gasoline sold in California will not be covered because California has already implemented more stringent standards similar to those EPA is establishing.)

1.Control of Hazardous Air Pollutants from Mobile Sources: Final Rule to Reduce Mobile Source Air Toxics, *United States Environmental Protection Agency*, EPA420-F-07-017, Feb 2007.

3

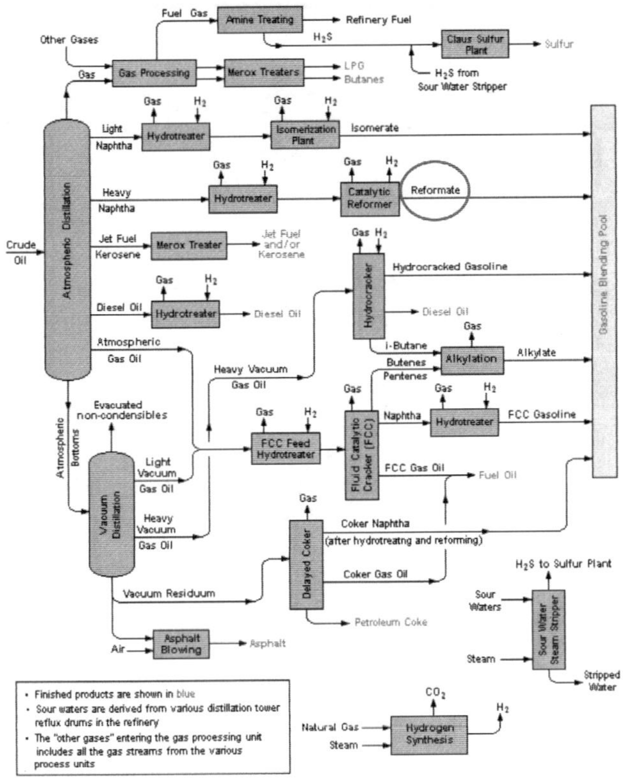

Gasoline Specifications

Octane Number (RON) : 89+ , 92+ , 95+

Benzene: < 0.62% vol

Gasoline Pool Dynamics[2]

	RON	%vol Bz
Light Naphtha	53 - 73	0.5 - 1.5
Reformate	91 - 104	4 - 9
Alkylate	91 - 96	0
FCC Gasoline	80 - 90	0.8 - 1.5

Benzene Pool[2]

	%vol Pool	%vol Bz	%Pool Bz	%Pool RON
Naphtha	11	0.6	3.1	8.8
Reformate	38	**4.5**	**80.7**	40.1
Alkylate	8	0	0	8.4
FCC	43	0.8	16.2	42.6

2. Coldwell, R. F. (2008). Benzene in gasoline, Regulation & remedies. *Process. Engineering Associates LLC.*

4

Benzene Removal

Catalytic Hydrogenation of Benzene

C_6H_6 *(Benzene)* + $3H_2$ ⟶ C_6H_{12} *(Cyclohexane)*

100 RON *83 RON*

C_7H_8 *(Toluene)* + $3H_2$ ⟶ C_7H_{14} *(Methylcyclohexane)*

120 RON *75 RON*

- Catalyst used for benzene hydrogenation also reduces toluene
- Toluene has high octane rating and needs to be retained in gasoline pool
- Capitalize on differing volatilities

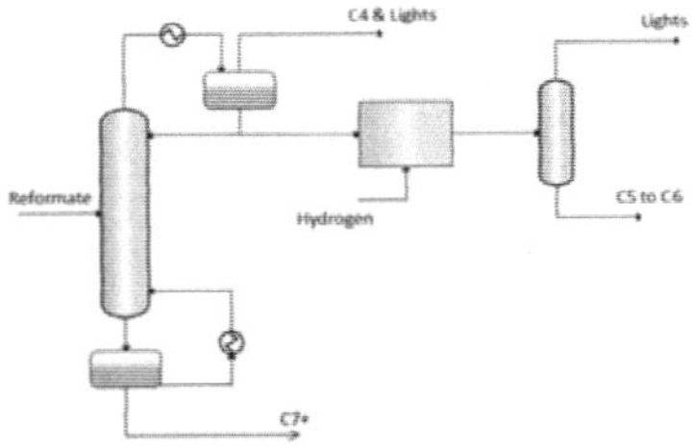

Conventional Process

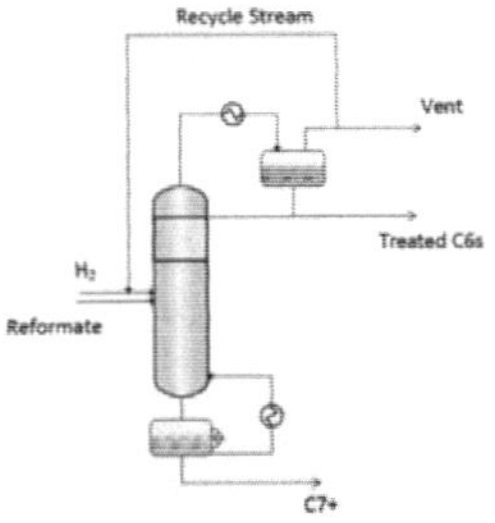

Reactive Distillation

Background – Reactive Distillation

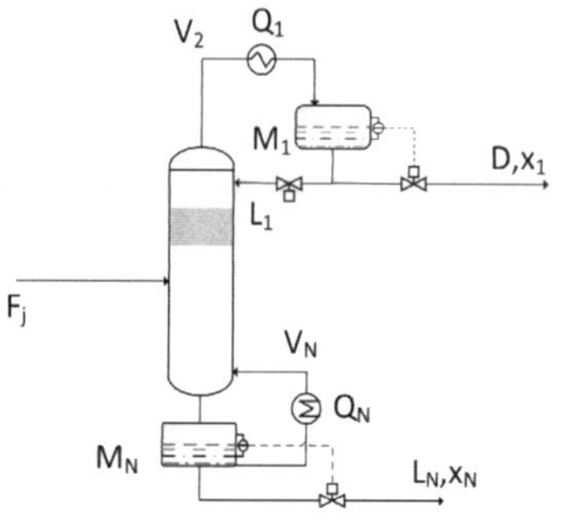

- Extension of conventional distillation
- Advantages
  - Lower capital investment
  - Heat of reaction can be directly used for separation
  - Reactive distillation columns have been found to be less susceptible to runaway reactions[3]
  - Can make use of different volatilities for reaction selectivity
- Applications: Desulfurization, Hydrogenation, Oligomerization, Isomerization

3. Harmsen, G. J. (2007). *Chemical Engineering and Processing: Process Intensification*, *46*(9), 774-780.

Challenges

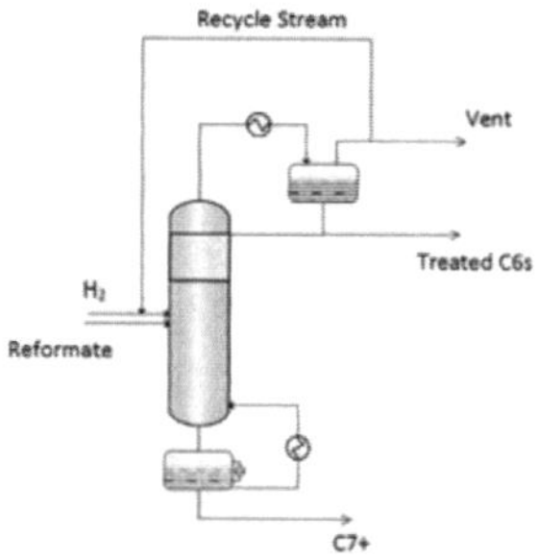

- Simultaneous presence of reaction and separation results in complex dynamic behavior
- Combining reaction and separation in one vessel results in fewer manipulated variables, increased interaction between control loops
- Feed composition of some components can vary by up to 50% (3–11 vol%)[4] and changes happen on a daily basis
- There is no literature available on modeling of benzene hydrogenation processes(!) and model has to be built from scratch
- Presence of partial condenser and recycle stream pose additional challenges to the model

Need to develop a rigorous dynamic model of the column and determine an effective control structure

4. Control of Hazardous Air Pollutants from Mobile Sources: 40 CFR Parts 59,50,85 and 86, *Environmental Protection Agency*, EPA-HQ-OAR-2005-0036, Feb 2006.

Column Design Parameters

- Packed Column approx. to staged column
- Light gases present, needs a Partial Condenser
- 70 stages, 15 components
- Dynamic Liquid and Vapor Holdup
- 2400 differential eqns, 5000 algebraic eqns
- HETP of 0.45m, column height : 30.6m
- Column diameter : 2.8m
- Standard 25mm Pall Rings packing assumed
- Equilibrium based kinetics
- Murphree efficiency of 0.9 used
- gPROMS used for development of model
- Peng Robinson Method used for VLE eqns
- MULTIFLASH used for all chemical properties

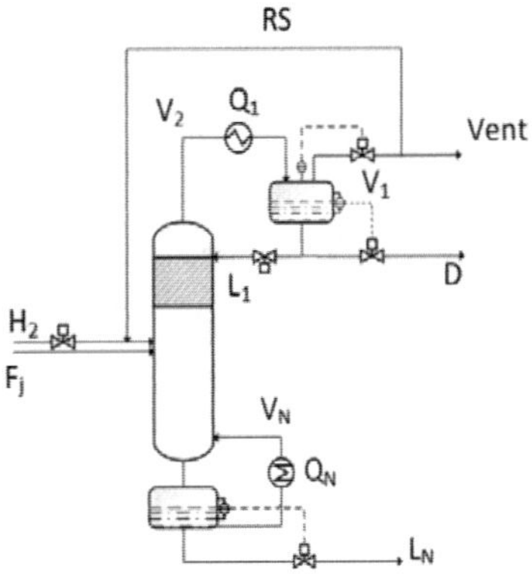

Model Development

Condenser & Reflux Drum

Mass Balance

$$\frac{d(N_{l_1} + N_{v_1})}{dt} = V_2 - L_1 - D - V_1$$

Component Balances

$$\frac{d(N_{l_1} x_{1,i} + N_{v_1} y_{1,i})}{dt} = V_2 y_{2,i} - L_1 x_{1,i} - V_1 y_{1,i} \qquad \forall\, i : 1\ to\ n-1$$

$$\sum_{i=1}^{n} x_{1,i} = 1 \qquad \sum_{i=1}^{n} y_{1,i} = 1$$

Vapor – Liquid Equilibrium

$$x_{1,i} \emptyset_{l_{1,i}} = y_{1,i} \emptyset_{v_{1,i}} \qquad \forall\, i : 1\ to\ n$$

Energy Balance

$$0 = V_2 H_{v_2} - V_1 H_0 + Q_1$$

$$\frac{d(N_{l_1} H_{l_1} + N_{v_1} H_{v_1})}{dt} = V_2 H_0 - (L_1 + D) H_{l_1} - V_1 H_{v_1}$$

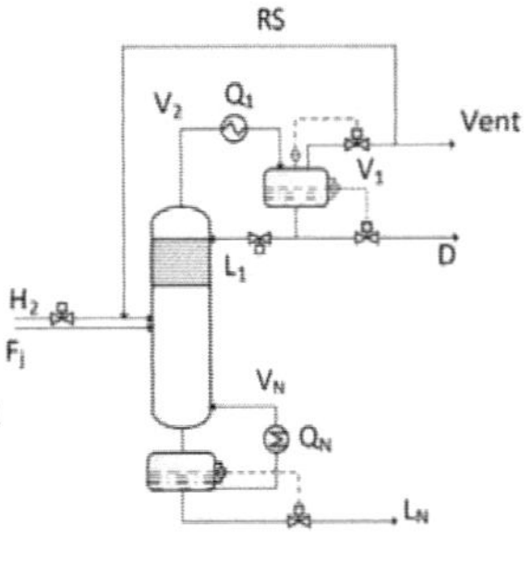

Model Development

Packed Section

Plate j

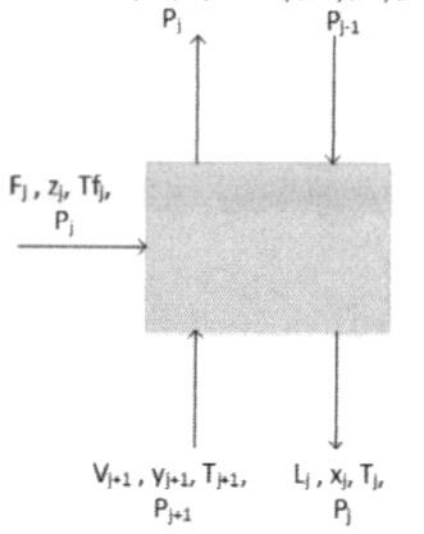

Mass Balance

$$\frac{d\left(N_{l_j} + N_{v_j}\right)}{dt} = F_j + V_{j+1} + L_{j-1} - V_j - L_j + Ah_j(HETP)\rho_{cat}\sum_{i=1}^{n} rxn_{j,i}$$

Component Balances

$$\frac{d\left(N_{l_j}x_{j,i} + N_{v_j}y_{j,i}\right)}{dt} = F_j z_{j,i} + V_{j+1}y_{j+1,i} - L_{j-1}x_{j-1,i} - V_j y_{j,i} - L_j x_{j,i} + Ah_j(HETP)\rho_{cat} rxn_{j,i} \quad \forall\, i : 1\ to\ n-1$$

$$\sum_{i=1}^{n} x_{j,i} = 1 \quad ; \quad \sum_{i=1}^{n} y_{j,i} = 1$$

Vapor – Liquid Equilibrium & Murphree Efficiency

$$x_{j,i}\emptyset_{l_{j,i}} = y^*_{j,i}\emptyset_{l_{j,i}} \qquad \forall\, i : 1\ to\ n$$

$$y_{j,i} = y_{j+1,i}(1-\eta) + y^*_{j,i}\eta \qquad \forall\, i : 1\ to\ n$$

Energy Balance

$$\frac{d\left(N_{l_j}H_{l_j} + N_{v_j}H_{v_j}\right)}{dt} = V_{j+1}H_{j+1} + L_{j-1}H_{l_{j-1}} - L_j H_{l_j} - V_j H_{v_j} - H_{rxn,j}$$

Model Development

Flow rate & Holdups

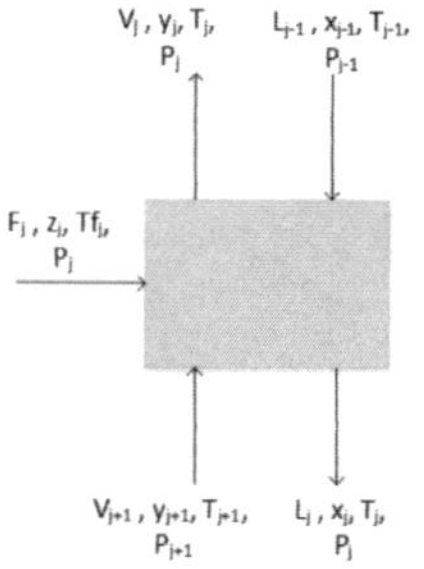

Liquid:

$$N_j vol_{l_j} = h_j(HETP)A$$

$$h_j = 0.34a^{1/3}u_j^{2/3}$$

Vapor:

$$P_j A(HETP)(\varepsilon - h_j) = N_{v_j} R_{gas} T_j$$

$$\frac{\Delta P_{0,j}}{HETP} = \psi\frac{(1-\varepsilon)}{\varepsilon^3}\frac{u_v^2\rho_v}{d_p K}$$

$$\Psi = \frac{150}{Re_v} + 1.75$$

$$\Delta P_j = \Delta P_{0,j}\left[1 - \frac{2h_j}{d_p a}\right]^{-5}$$

$$P_{j+1} = P_j + \Delta P_j$$

Reaction Rate[5]

$$rxn_{j,Benzene} = -\frac{k_{j,1}K_{A_1}K_{H_1}C_{A_{j,1}}C_{H_j}}{\left(3K_{A_1}C_{A_{j,1}} + \left(K_{H_1}C_{H_j}\right)^{\frac{1}{2}} + 1\right)^3}$$

$$rxn_{j,Toluene} = -\frac{k_{j,2}K_{A_2}K_{H_2}C_{A_{j,2}}C_{H_j}}{\left(3K_{A_2}C_{A_{j,2}} + \left(K_{H_2}C_{H_j}\right)^{\frac{1}{2}} + 1\right)^3}$$

$$k_{i,j} = k_{i,0}exp\left[-\frac{E_a}{R_{gas}}\left(\frac{1}{T_j} - \frac{1}{T_0}\right)\right]$$

	Benzene	Toluene
$k_{i,0}$ $mol/(skg)$	1.7 ± 0.8	3.0 ± 1
E_a kJ/mol	37.7 ± 1.9	16.8 ± 0.9
$K_{A_i}\ x\ 10^5$ m^3/mol	11.9 ± 4.1	11.9 ± 1.8
$K_{H_i}\ x\ 10^4$ m^3/mol	74.3 ± 69.1	21.9 ± 11.4

5. Toppinen, S. et al.; Ind. Eng. Chem. Res. 1996, 35, 1824-1833

Model Development

Reboiler

Mass Balance

$$\frac{d(N_{l_N})}{dt} = L_{N-1} - L_N - V_N$$

Component Balances

$$\frac{d(N_{l_N} x_{N,i})}{dt} = L_{N-1} x_{N-1,i} - L_N x_{N,i} - V_N y_{N,i} \qquad \forall\ i : 1\ to\ n-1$$

$$\sum_{i=1}^{n} x_{N,i} = 1$$

$$x_{N,i} = y_{N,i} \qquad \forall\ i : 1\ to\ n$$

Energy Balance

$$\frac{d(N_{l_N} H_{l_N})}{dt} = L_{N-1} H_{l_{N-1}} - (L_N + V_N) H_{l_N}$$

$$V_N H_{l_N} - V_N H_{v_N} + Q_N = 0$$

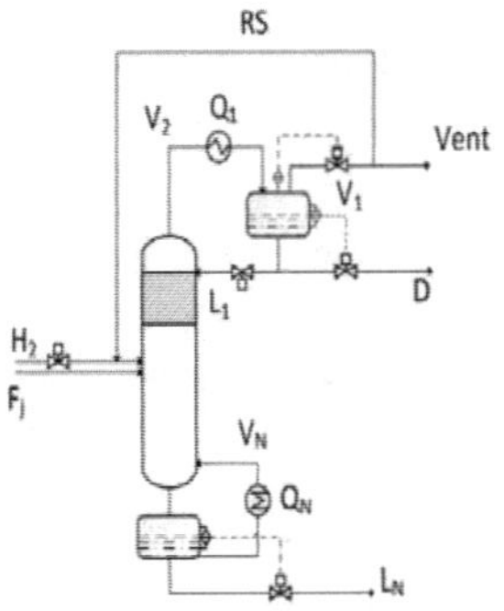

Steady State Parameters

- Feed flow rate : 265 mols^{-1} / 91400 kghr^{-1}
- H_2 flow rate : 65.9 mols^{-1} / 1158.4 kghr^{-1}
- Feed temp : 430 K
- Reflux ratio : 4.31
- Distillate temp : 293.5 K
- Bottom temp : 495.5 K
- Column Pressure : 800 kPa
- Number of catalyst stages : 10
- Condenser Duty : -2.102E7 Js^{-1}
- Reboiler Duty : 1.567E7 Js^{-1}
- Benzene in feed : 6.01 vol%
- Bz product : 0.32 vol%, 1.26wt%
- Toluene retained : 99.85%

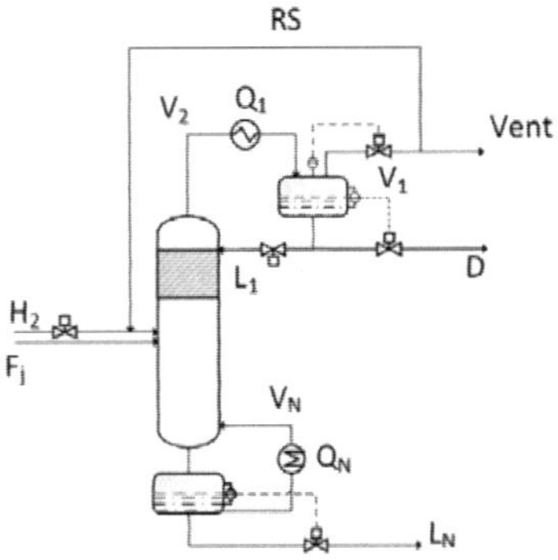

Manipulated & Controlled Variables

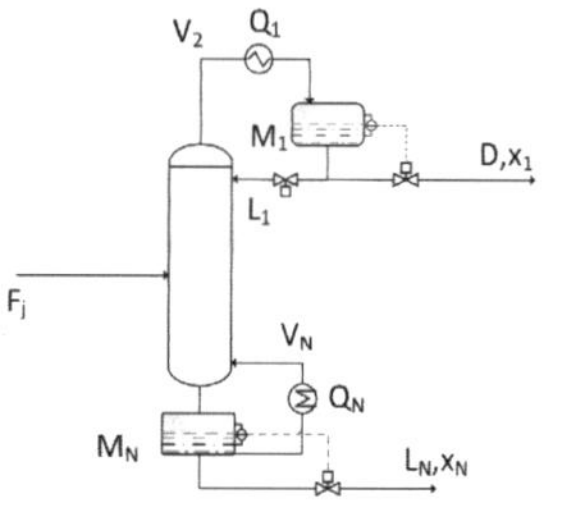

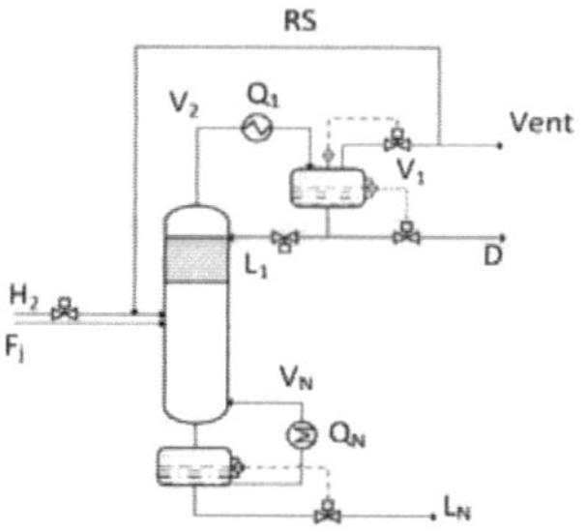

Common Control Configurations
4x4 or 5x5 control[6-8]

Due to presence of partial condenser, and recycle stream there are additional degrees of freedom

Manipulated Inputs (u)

- Distillate Flow, D
- Bottom Flow, L_N
- Reflux Ratio, R
- Reboiler Duty, Q_N
- Condenser Duty, Q_c

Controlled Outputs (y)

- Reflux drum liquid level, h_1
- Reboiler liquid level, h_N
- Distillate product composition, x_1 — T_1
- Bottom product composition, x_N — T_{55}
- Pressure at top of column, P_1

6. Skogestad, S., et al. (1987). *AIChE Journal, 33*(10), 1620-1635. 7. Skogestad, S., et al. (1990). *AIChE Journal, 36*(5), 753-764.
8. Skogestad, S. (1997). *Modeling, Identification and Control, 18*(3), 177-217.

Manipulated & Controlled Variables

$$u = \begin{pmatrix} L_1 \\ L_N \\ R \\ Q_N \\ Q_1 \\ V_1 \\ F_{H2} \end{pmatrix}$$

$$\begin{matrix} u & & y \end{matrix}$$
$$\begin{pmatrix} L_1 & \leftrightarrow & h_1 \\ V_1 & \leftrightarrow & P_1 \\ L_N & \leftrightarrow & h_N \\ F_{H_2} & \leftrightarrow & V_1 \end{pmatrix}$$

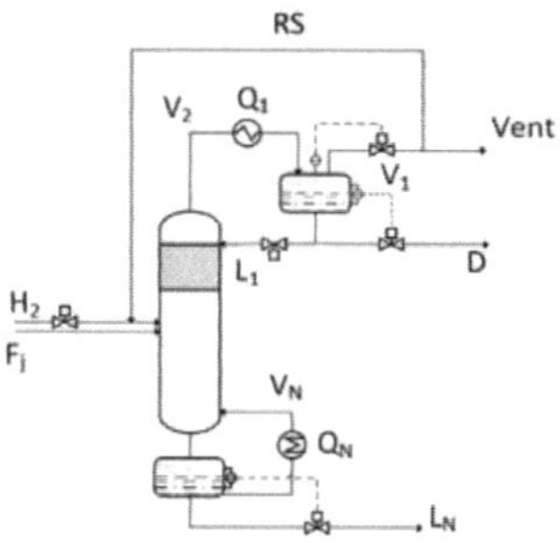

$$\overset{u}{\begin{pmatrix} Q_1 \\ Q_N \\ R \end{pmatrix}} \quad \overset{y}{\begin{pmatrix} T_1 \\ T_{19} \\ T_{55} \end{pmatrix}}$$

- Need to select one more controlled variable
  - Avoid trays near the feed tray
  - Avoid trays near the top or the bottom of the column
  - Avoid the catalyst zone
- Need to pair the remaining 3 sets of variables based on their sensitivity
- This can be done using Relative Gain Array (RGA) analysis

15

Pairing of variables

	Q_1	Q_N	R
T_1	$\frac{1.0684x10^{-4}}{1+1887.1s}$	$\frac{1.165x10^{-4}(1+6052.2s)}{1+2(0.815)(3850.5s)+(3850.5s)^2}$	$\frac{-36.523(1+9632.8s)}{1+2(0.901)(3548.8)s+(3548.8s)^2}$
T_{19}	$\frac{1.8704x10^{-5}}{(1+4555.3s)(1+1834.3s)}$	$\frac{2.524x10^{-5}}{1+5817.4s}$	$\frac{-15.972}{1+4565.7s}$
T_{55}	$\frac{1.21x10^{-5}}{1+2(0.923)(3013.9)s+(3013.9s)^2}$	$\frac{1.4123x10^{-5}}{(1+3081.3s)(1+2259.4s)}$	$\frac{-10.502}{1+3593.2s}$

RGA Matrix

	Q_1	Q_N	R
T_1	1.800	−0.157	−0.643
T_{19}	−5.646	7.322	−0.676
T_{55}	4.846	−6.165	2.319

$$\begin{matrix} u & & y \end{matrix}$$
$$\begin{pmatrix} L_1 & \leftrightarrow & h_1 \\ V_1 & \leftrightarrow & P_1 \\ L_N & \leftrightarrow & h_N \\ F_{H_2} & \leftrightarrow & V_1 \\ Q_1 & \leftrightarrow & T_1 \\ Q_N & \leftrightarrow & T_{19} \\ R & \leftrightarrow & T_{55} \end{pmatrix}$$

16

Controller Parameters

Manipulated Variable	Controlled variable	K_c	τ_I (sec)	τ_c (sec)	Type of Controller and Tuning Method Used
Q_1	T_1	$\frac{1.766x10^7}{\tau_c}$	1887.1	110	PI (IMC)
Q_N	T_{19}	$\frac{2.305x10^8}{\tau_c}$	5817.4	582	PI (IMC)
R	T_{55}	$-\frac{342.15}{\tau_c}$	3593.3	610	PI (IMC)
F_{H_2}	V_1	$\frac{9.013}{\tau_c}$	678.4	128.7	PI (IMC)
D	h_1	3340	-	-	P (Ziegler Nichols)
L_N	h_N	3830	-	-	P (Ziegler Nichols)
V_1	P_1	12.5	-	-	P (Ziegler Nichols)

Response to Step Disturbances

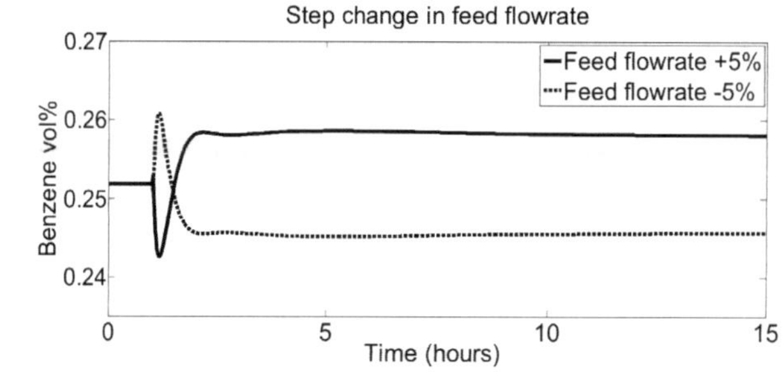

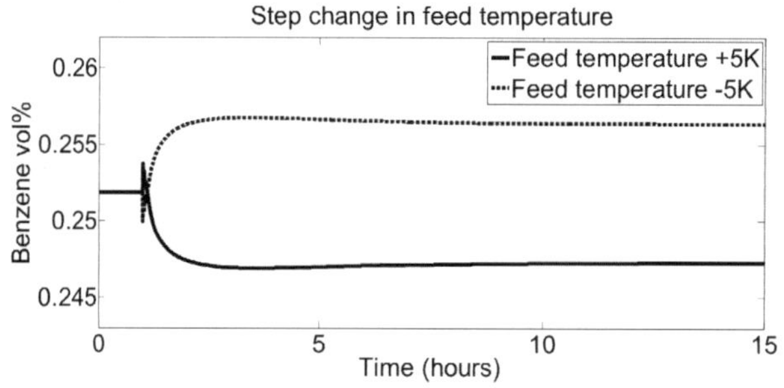

Response to Step Disturbances in Feed Composition

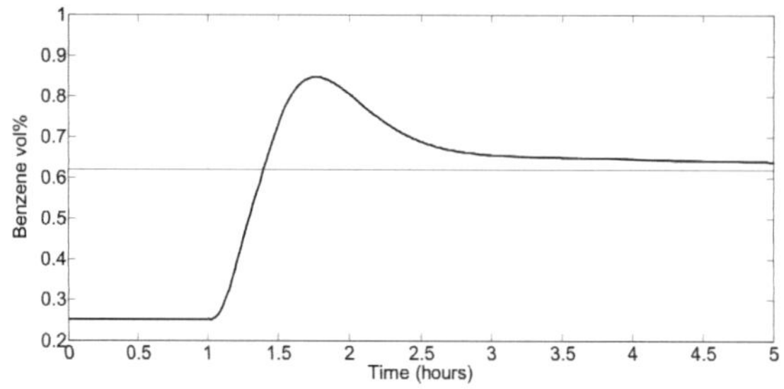

Feedback only control does not take a corrective action until after a deviation in the controlled variables occurs.

Large overshoot and long settling time for feed disturbances on the product streams suggests to investigate predictive control action

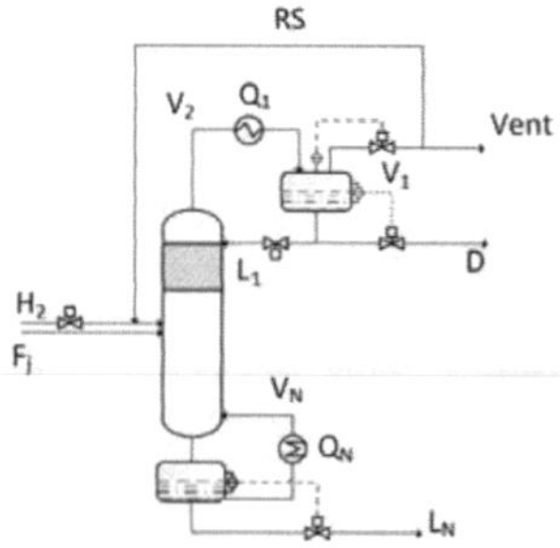

Feedforward Control

Feedforward controller design

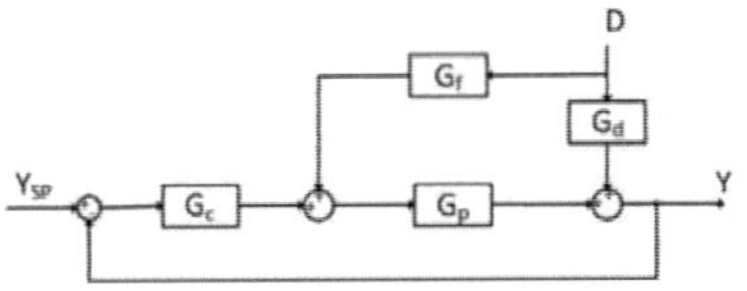

Block diagram of a feedforward control system

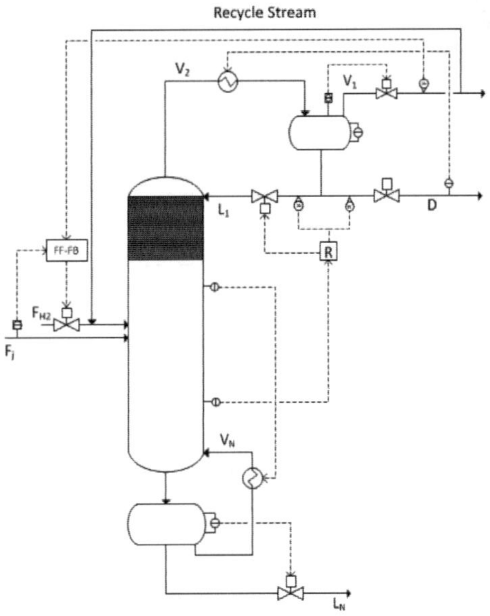

$$G_f = -\frac{G_d}{G_p}$$

$$G_p = \frac{V_1}{F_{H_2}} \quad ; \quad G_d = \frac{V_1}{z_{C_6H_6}}$$

$$G_p = \frac{99.39}{1 + 3269.9s}$$

$$G_d = \frac{-7.457x10^4}{1 + 3580.2s}$$

Feedforward-Feedback Control

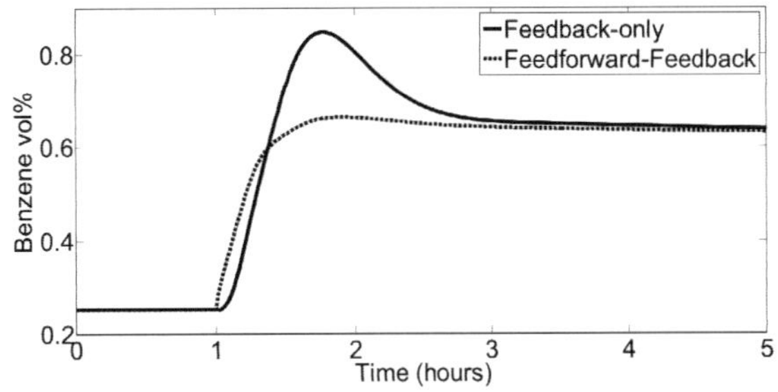

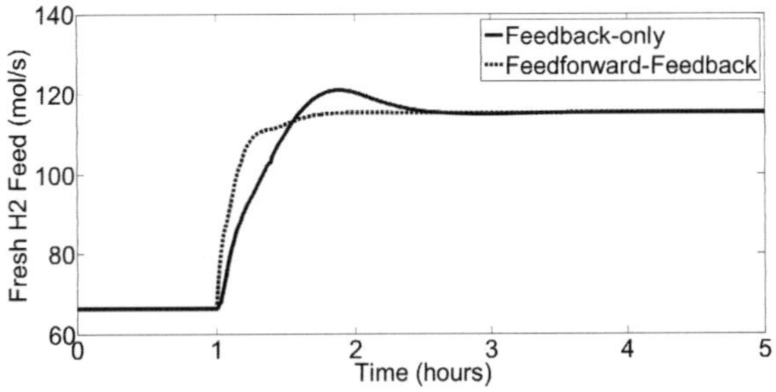

Response to Step Disturbances in Feed Composition

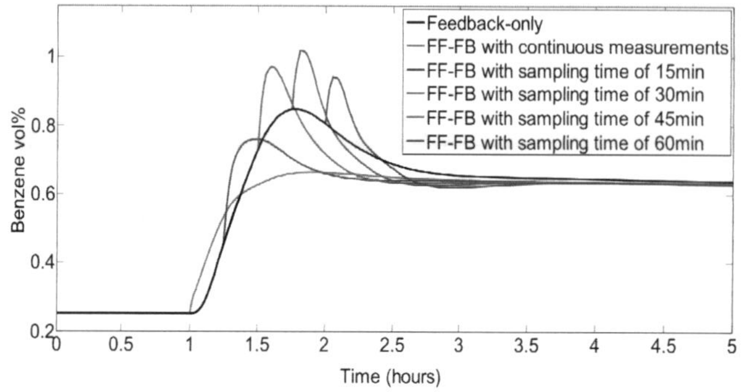

Control structure	*Dev*	% reduct.	Max Bz vol%
Feedback-only	1074		0.85
FF-FB with continuous measurements	423	61%	0.66
FF-FB with sampling time of 15min	559	48%	0.76
FF-FB with sampling time of 30min	756	30%	0.97
FF-FB with sampling time of 45min	874	19%	1.02
FF-FB with sampling time of 60min	917	15%	0.94

Conclusion

- Column performance for feed temperature and flow rate disturbances remains acceptable for feedback-only control scheme
- Column has a significant transition and settling time for disturbances in feed composition
- Feedforward control reduces upset conditions resulting from feed composition disturbances. However, this is only beneficial if there is little time delay associated with the feed composition measurement.

23

Acknowledgements

- Financial support
  - NSF-CDI (CBET#0941313)
  - ACS (PRF#50978-ND9)

24

Thank you

Combining Process Simulation and CFD Simulation to Improve the Efficiency of SMR (Steam Methane Reformer Plants).

Fan, R., Air Liquide

Originally presented on: 6/11/2014 10:00:00 - 10:25:00

Steam methane reforming is an endothermic process carried out in nickel-based catalyst filled tubes inside a large gas fired furnace. Due to the high temperature of the combustion gas and the furnace walls, radiation is the predominant mode of heat transfer in the furnace. The furnace walls in the tubular reformer are lined with refractory, bricks and fiber materials having a relative low emissivity. By applying a "high emissivity coating" on the surface of the wall, it is possible to increase the heat transferred to the tubes (mainly radiation) and thereby improve the thermal efficiency of the furnace box. Considering the scale and the important of the industrial process even a small rise in the heat transfer to the tubes can be translated into an important increase in the hydrogen production rate or an important decrease in the required fuel consumption [1]. In this article, the impact of high emissivity coating on the global efficiency of a side-fired SMR furnace was conducted through process simulation (Aspen HYSYS) and 3D CFD simulation.

Process Engineering for Environmental footprinting

Shaibal Roy, Stephen Tieri, Todd Krieger

Process Development Symposium, Philadelphia
12th June 2014

The miracles of science™

Overview

Background – Environmental evaluation

Life Cycle Assessment (LCA) – a tool for Environmental evaluation

LCA methodology and tools

LCA applications and examples

Summary

Any good process design should include

- Energy Conservation
- Reduced Greenhouse Gas Emissions
- Water Conservation
- Reduced Material Use
- Renewable Sources
- Reduced Waste
- Simplified Disposal
- Less Pollution
- Less Impact on Ecosystems
- Reduced Toxicological Risk

LCA Integrated with Process Development

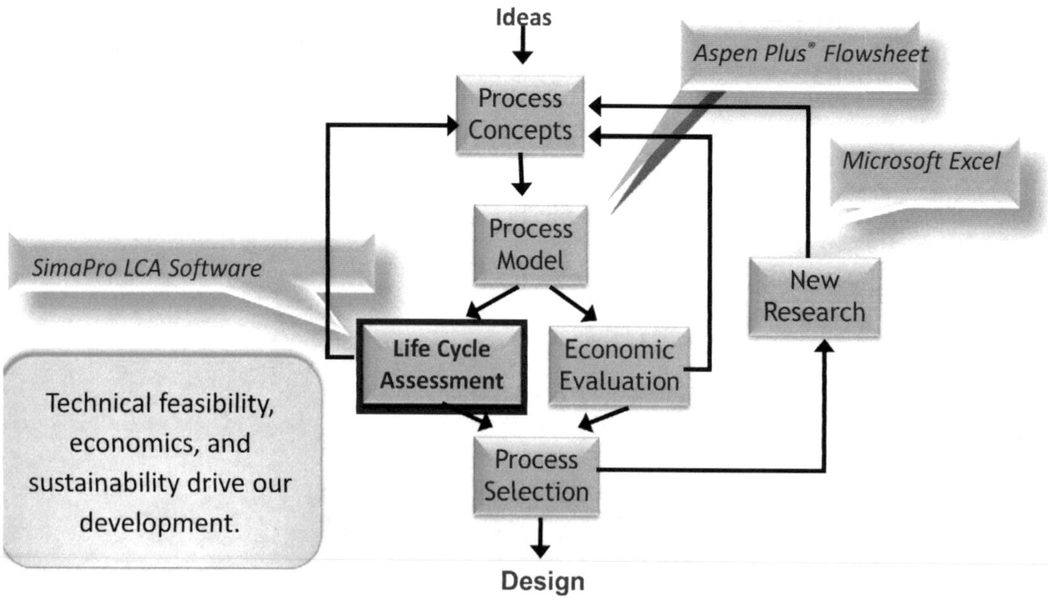

C. M. Alles , R. Jenkins, Sustainable Biorefineries
http://books.nap.edu/openbook.php?record_id=12027&page=75

What is meant by Life Cycle Analysis?

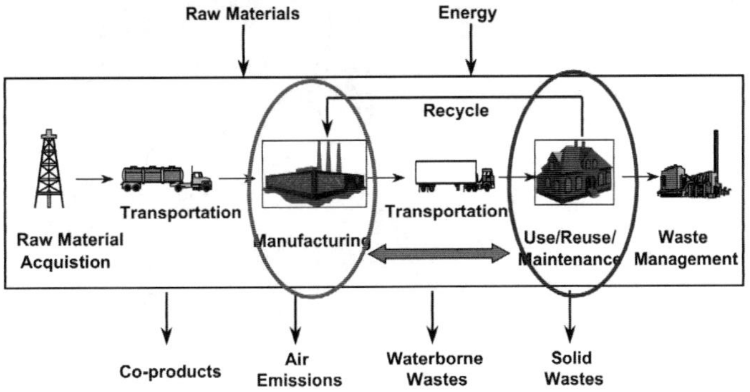

Higher energy consumption of plastics in the production phase

Energy savings in use: - longer life span
- superior functionality

Holistic approach: evaluate the environmental impact of a product system, including all material and energy flows along the entire supply chain

LCA is the only internationally recognized standard method to evaluate the environmental footprint of entire supply chains.

=> ISO standards 14040 cc.

LCA Methodology (ISO standards 14040 cc)

1. **Goal and Scoping**
   - LCA practitioner and all other stake holder develop system boundary and define functional unit
   - Efficacy and efficiency of the study depends on this step
2. **Life Cycle Inventory (LCI)**
   - Process calculations to estimate unit consumptions, data collection from internal and external sources
   - Often the most time consuming step
3. **Life Cycle Impact Assessment (LCIA)**
   - Standard assessment embedded in LCA software
4. **Interpretation / Customer communication**
   - Needs sensitivity study, often lead to revisions
   - requires case-by-case peer review of data and methods

Example: Foam Expansion Agent

Spray polyurethane foam (SPF) insulation

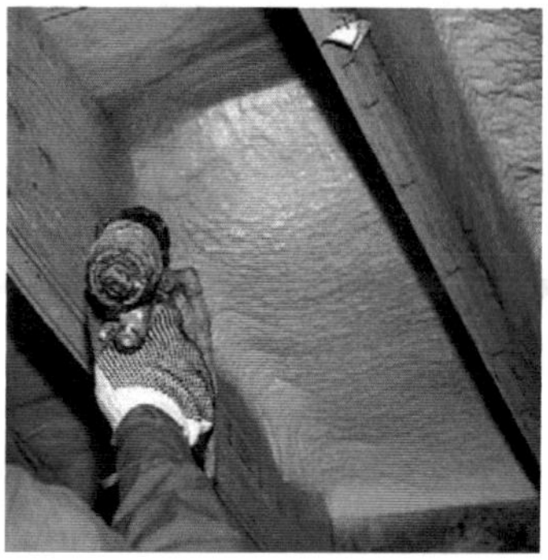

Mixed at location

The "A" side	The "B" side
Isocynate	Polyol
	Surfactant
	Flame retardant
	Catalyst
	Foam expansion agent

Foam expansion agent →
- HFO-1336mzz-Z
- HCFC-141b
- HFC-245fa
- H_2O/CO_2
- Other historical FEAs

Environmental concerns – the tradeoffs

Concerns about climate change of current commercial FEAs - creating demand for products with low greenhouse gas footprints

Global Warming Potential (GWP) for Foam Expansion Agents		
Foam expansion agent	**Global Warming Potential, kg CO_2 eq per kg FEA**	**References**
HCFC-141b	725	[IPCC 4th ed. 2007]
HFC-245fa	1030	[IPCC 4th ed. 2007]
HFO-1336mzz-Z	8.9	[Baasondorj, M., et.al]
Hydrocarbon	11	[IPCC 4th ed. 2007]

In addition, the GHG due to

1. Foam formulation Manufacturing
2. Loss during installation
3. Emission during use
4. End of life
5. **Energy savings during use**

A holistic life cycle approach is necessary

LCA Methodology

1) **Goal and Scoping**
2) Life Cycle Inventory (LCI)
3) Life Cycle Impact Assessment (LCIA)
4) Life Cycle Interpretation

Goal and Scoping

Cradle-to-Grave System Boundaries

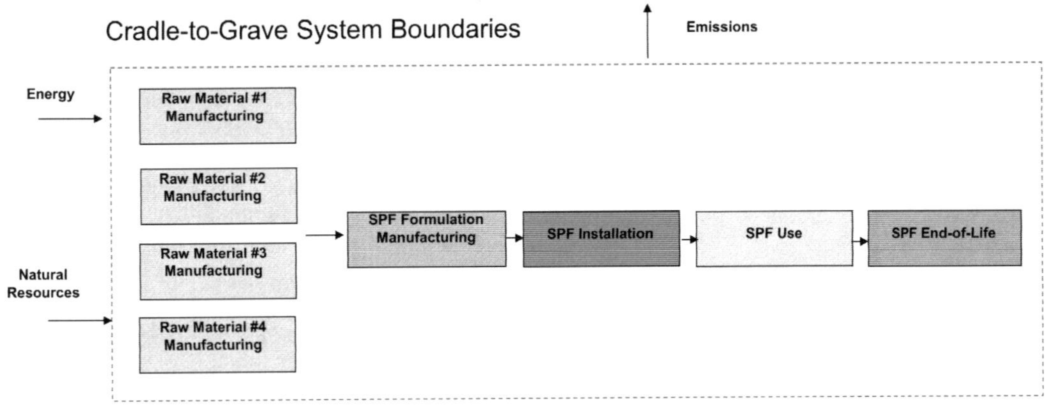

- Functional unit (application based): one square-meter of foam insulation in a residential unit
  - At constant **R** value
  - At constant thickness
- Geography: Baltimore, Chicago, Houston
- Primary impact: Climate Change Potential (CCP) (as measured by kg CO_2 equivalents)

LCA Methodology

1) Goal and Scoping
2) **Life Cycle Inventory (LCI)**
3) Life Cycle Impact Assessment (LCIA)
4) Life Cycle Interpretation

Life Cycle Inventory (LCI)

- Data collected from formulators
  - Raw material inputs (Polyol, FEA, Isocyanate, Catalyst, Fire Retardant, Surfactant) from process calculation, published information
  - Performance inputs (R-value per inch) from experiments
- FEA Emissions during Install, Use, D&R, EOL from published information
- Energy Use in the house from design software (e.g. Trane®)
- Other inventory data from software database

FEA mass balance (Example)

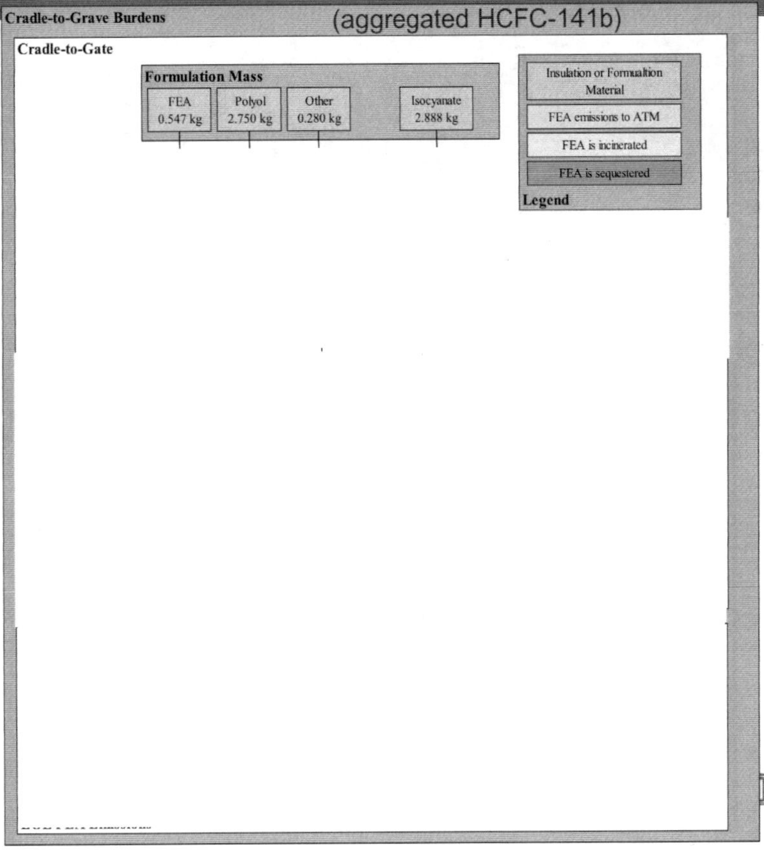

EOL Assumptions

Incineration / Landfill Split
- **based on US MSW data (17% / 83%)**

Incineration - FEA
- **destroyed, but include combustion emissions of FEA & insulation**

Landfill - Assume 50% FEA
- **is sequestered beyond 100-yr temporal boundary of study, 50% emitted to atm.**

Simapro-HCFC 141b

Products

Known outputs to technosphere. Products and co-products

Name	Amount	Unit	Quantity	Allocation %	Waste type	Category
HCFC-141b FOR DKC ILLU	1	lb	Mass	100 %	not defined	...\Void - or 2011 paper
(Insert line here)						

Known outputs to technosphere. Avoided products

Name	Amount	Unit	Distribution	SD^2 or 2*SD	Min	Max
Chlorine, liquid, production mix, at plant/US WITH US ELECTRICITY DUPONT U	0.30	lb	Undefined			
(Insert line here)						

Inputs

Known inputs from nature (resources)

Name	Sub-compartment	Amount	Unit	Distribution	SD^2 or 2*SD	Min	Max	Comment
(Insert line here)								

Known inputs from technosphere (materials/fuels)

Name	Amount	Unit	Distribution	SD^2 or 2*SD	Min
Methyl Chloroform (SRI) US 2012	1.14	lb	Undefined		
Hydrogen fluoride, at plant/US DUPONT (LP) U	0.17	lb	Undefined		
Sodium hydroxide, 50% in H2O, production mix, at plant/RER WITH US ELECTRICITY DUPON	0.03	lb	Undefined		
Transport, transoceanic freight ship/OCE U	0.1	kgkm	Undefined		
(Insert line here)					

Known inputs from technosphere (electricity/heat)

Name	Amount	Unit	Distribution	SD^2 or 2*SD	Min
Steam, NG indutrial boiler/RNA US-EI DUPONT 111611	0.83	lb	Undefined		
Electricity, medium voltage, at grid/US WITH US ELECTRICITY U	0.07	kWh	Undefined		

Input data available from plant energy and mass balance, Aspen plus, other published reports

Simapro-HCFC 141b

Outputs

Emissions to air

Name	Sub-compartment	Amount	Unit	Distribution	SD^2 or 2*SD	Min
Ethane, 1,1-dichloro-1-fluoro-, HCFC-141b	high. pop.	1.02E3	lb	Undefined		
Ethane, 1-chloro-1,1-difluoro-, HCFC-142b	high. pop.	7.55E-7	lb	Undefined		
Ethane, 1,1,1-trichloro-, HCFC-140	high. pop.	2.84E-5	lb	Undefined		
Ethane, 1,1,1-trifluoro-, HFC-143a	high. pop.	1.13E-6	lb	Undefined		
Hydrogen chloride	high. pop.	4.66E-5	lb	Undefined		
Hydrogen fluoride	high. pop.	1.14E-5	lb	Undefined		
Carbon dioxide, fossil	high. pop.	8.8E-3	lb	Undefined		
Ethane, 1,1,1-trichloro-, HCFC-140	high. pop.	1.14E-3	lb	Undefined		
(Insert line here)						

Emissions to water

Name	Sub-compartment	Amount	Unit	Distribution	SD^2 or 2*SD	Min	Max	Comment
(Insert line here)								

Emissions to soil

Name	Sub-compartment	Amount	Unit	Distribution	SD^2 or 2*SD	Min	Max	Comment
(Insert line here)								

Final waste flows

Name	Sub-compartment	Amount	Unit	Distribution	SD^2 or 2*SD	Min	Max
Sodium Chloride		1.57E-2	lb	Undefined			
Sodium Fluoride		5.05E-3	lb	Undefined			

Emissions known from plant data, govt. reported data, TRI etc.

LCA Methodology

1) Goal and Scoping
2) Life Cycle Inventory (LCI)
3) **Life Cycle Impact Assessment (LCIA)**
4) Life Cycle Interpretation

Life Cycle Impact Assessments

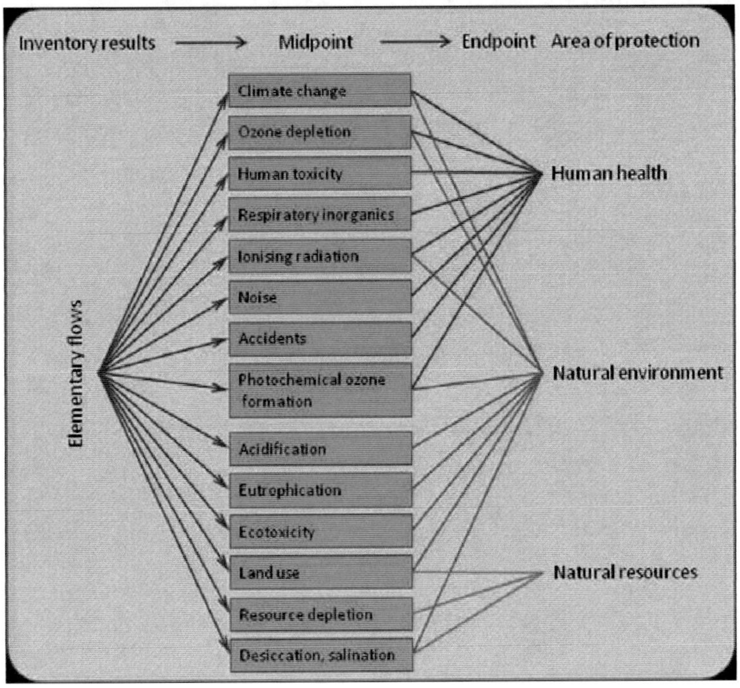

http://lct.jrc.ec.europa.eu/

Impact assessment results

Simapro-HCFC 141b

Sel	Impact category	Unit	Total	HCFC-141b FOR DKC	Methyl Chloroform	Hydrogen fluoride, at	Sodium hydroxide,	Transport, transoceanic	Steam, NG indutrial	Electricity, medium	Chlorine, liquid,	Process-spec burdens,
☑	Ozone depletion	kg CFC-11 eq	122	122	1.71E-5	2.49E-8	1.01E-9	3.29E-13	2.93E-8	5.27E-9	-5.03E-7	6.08E-12
☑	Global warming	kg CO2 eq	7.4E5	7.4E5	3.21	0.258	0.033	1.94E-6	0.173	0.118	-0.289	3.65E-5
☑	Smog	kg O3 eq	0.116	5.7E-6	0.0886	0.0272	0.00093	7.61E-7	0.00251	0.00527	-0.00873	1.16E-5
☑	Acidification	kg SO2 eq	0.0311	5.92E-5	0.0247	0.00745	0.000261	5.05E-8	0.000142	0.000795	-0.00227	4E-7
☑	Eutrophication	kg N eq	0.00444	0	0.00374	0.000713	5.34E-5	2.42E-9	1.49E-5	0.000444	-0.000526	3.21E-8
☑	Carcinogenics	CTUh	4.92E-5	4.92E-5	6.01E-8	8.44E-9	7.53E-10	3.31E-15	1.37E-10	5.73E-9	-7.33E-9	1.44E-13
☑	Non carcinogenics	CTUh	3.26E-7	2.22E-11	3.22E-7	2.44E-8	4.07E-9	3.84E-14	5.2E-10	1.37E-8	-3.86E-8	3.37E-13
☑	Respiratory effects	kg PM2.5 eq	0.00194	0	0.0015	0.000505	1.58E-5	3.02E-9	6.1E-6	4.81E-5	-0.000138	1.04E-7
☑	Ecotoxicity	CTUe	3.32	0.00024	3.21	0.202	0.0247	1.89E-7	0.00222	0.0982	-0.218	3.65E-6
☑	Non-Renewable depletion	kg oil eq	1.64	0	1.53	0.0918	0.013	6.83E-7	0.072	0.0449	-0.112	1.52E-5

- ❑ Global warming is not the only environmental impact
- ❑ Depending on the product and goal, other categories may become more important.
- ❑ The result can be normalized w.r.t. base value to gauge their relative importance

LCA Methodology

1) Goal and Scoping
2) Life Cycle Inventory (LCI)
3) Life Cycle Impact Assessment (LCIA)
4) **Life Cycle Interpretation**

Life Cycle Interpretation

- Evaluate data quality
- Identify significant issues
- Draw conclusions & recommendations
- Interpret and communicate results

Cradle-to-Grave Results:

R-Value Basis - R-13 – Houston

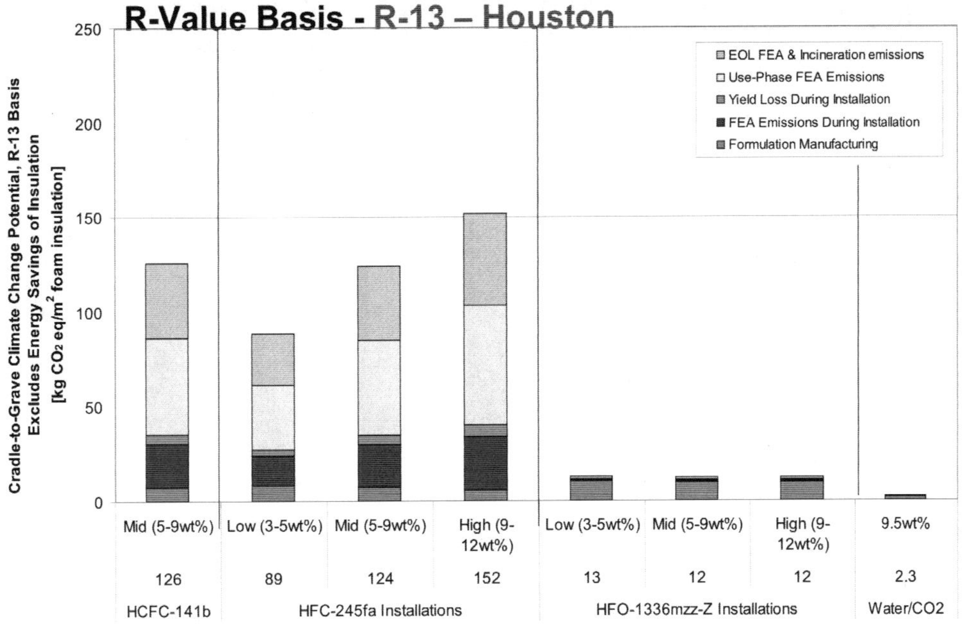

- Formulation manufacturing burden for HFO-1336 mzz-Z is slightly higher than other halogenated FEAs
- Due to low GWP of HFO, its emission during installation, use and disposal phase has significantly less burden among closed cell foams.

Cradle-to-Grave Results:

R-Value Basis - R-20 – Baltimore / Chicago

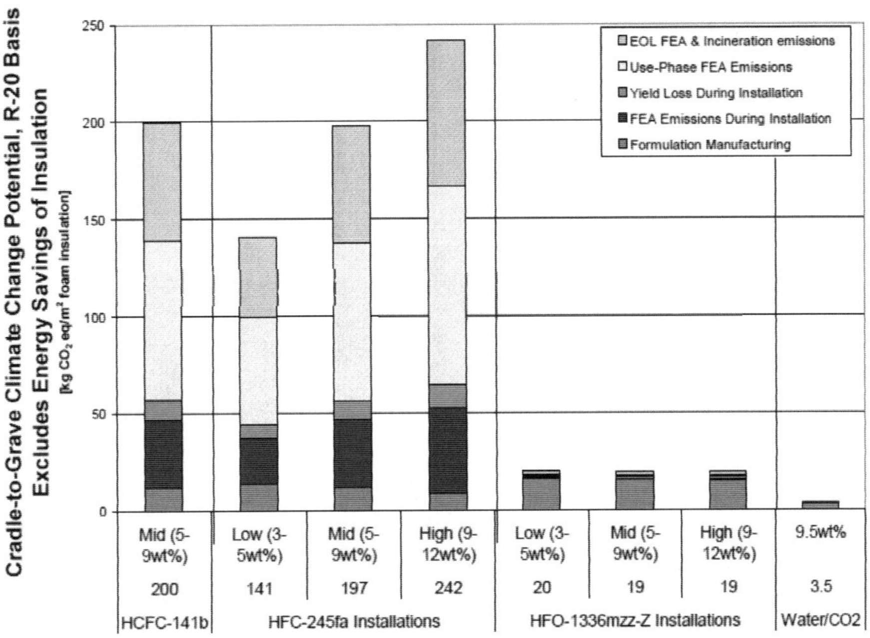

- ❑Higher burdens than in Houston due to colder climates and higher R value insulation
- ❑Similar relative burdens for each life cycle steps

Cradle-to-Grave Results:

Lifecycle GHG for 5.5" Basis - Chicago

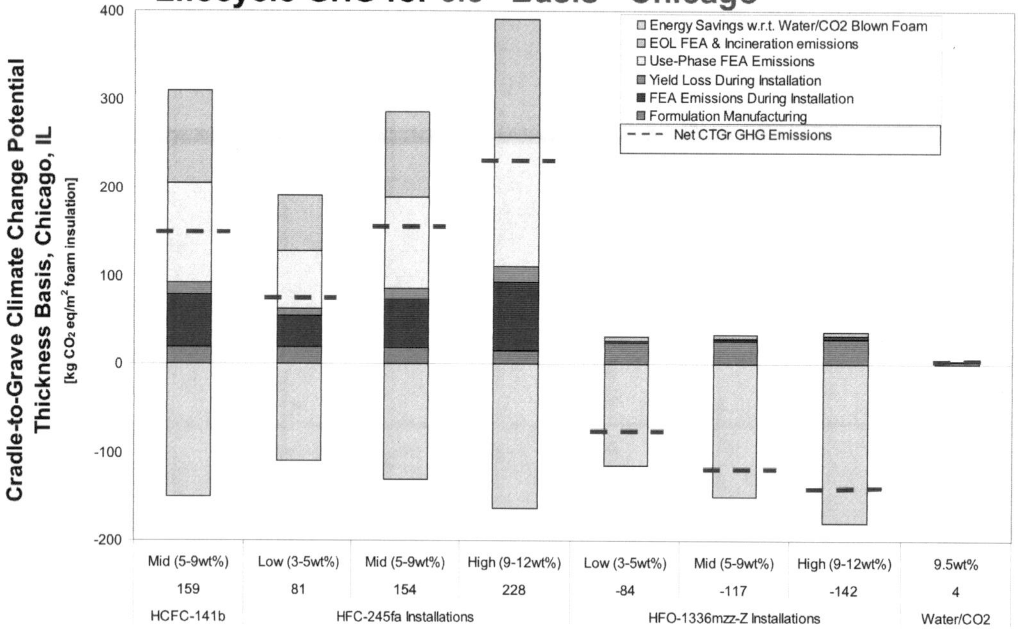

Increasing FEA results in higher energy savings and lower overall Climate Change Potential (CCP)

Sensitivity Analysis:

FEA Emission Assumptions

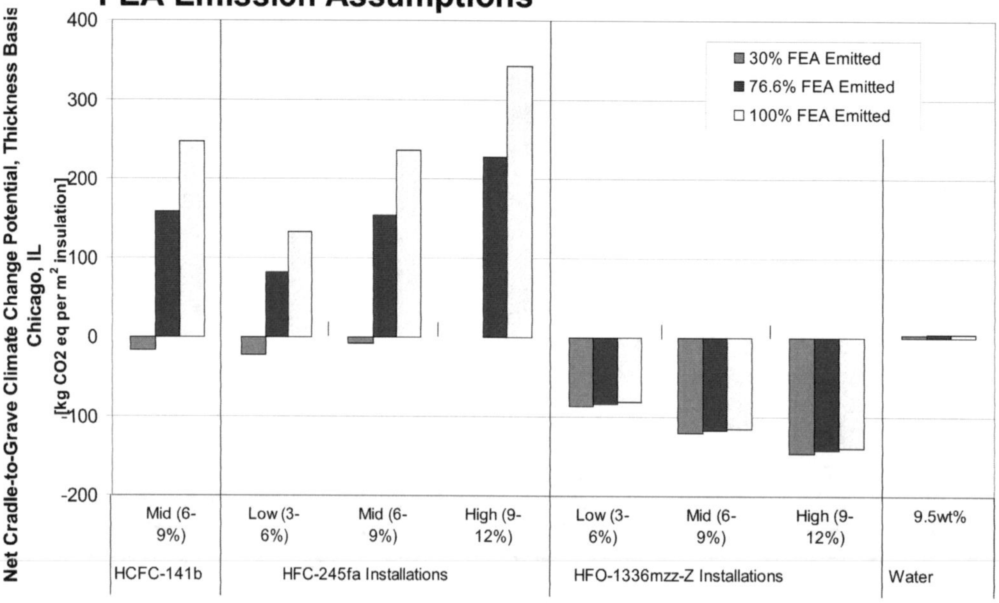

Net Climate Change potential for HFO-1336mzz-Z always lower

LCA Limitations

- local vs. global conditions
- aggregated emissions vs. threshold values
- no risk assessment
- traditionally focused on environmental vs. economic and societal value

Summary

- Life Cycle Assessment – a complementary analysis tool to get more insight into decision making
- Similar approach as in Techno-Economic evaluation, but with a broader system boundary
- Helps understand the impact of various steps in supply-chain of a process / product
- Identify and help understand the trade offs among different process options
- Sensitivity analysis to mitigate uncertainties in model assumptions

Questions?

Now, or later:

shaibal.roy@dupont.com

The miracles of science™

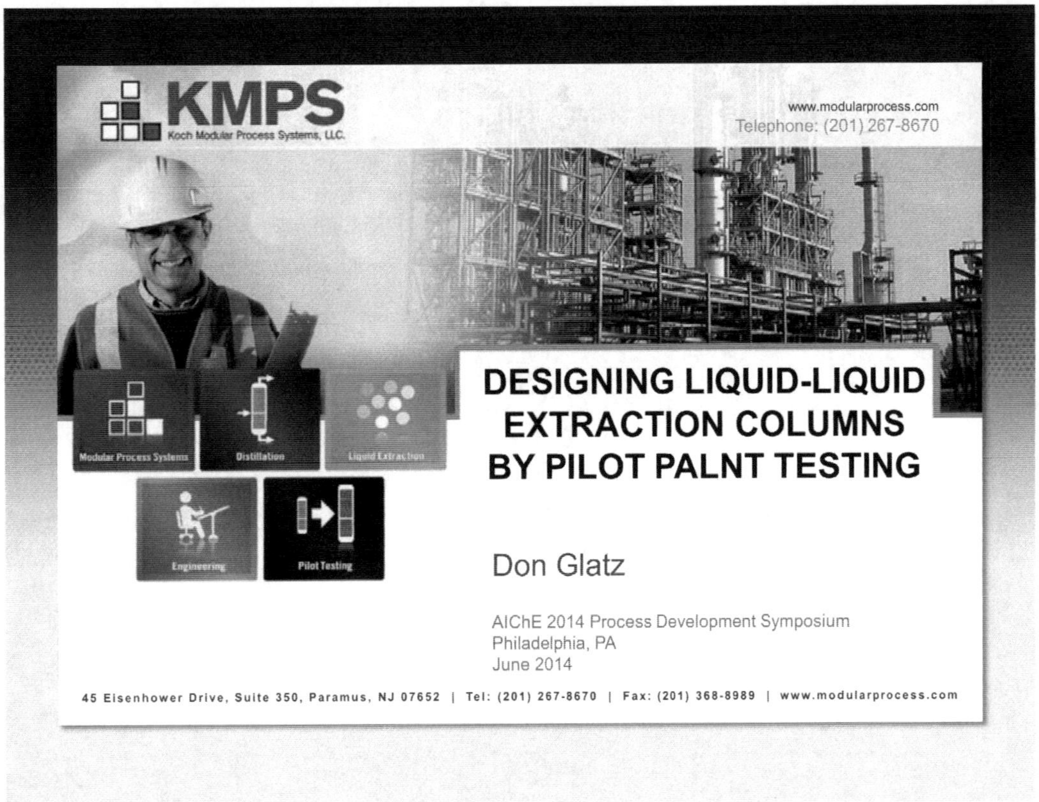

Simple Extraction Single Stage

KMPS Koch Modular Process Systems, LLC

MODULAR PROCESS SYSTEMS | DISTILLATION | LIQUID EXTRACTION | ENGINEERING | PILOT TESTING

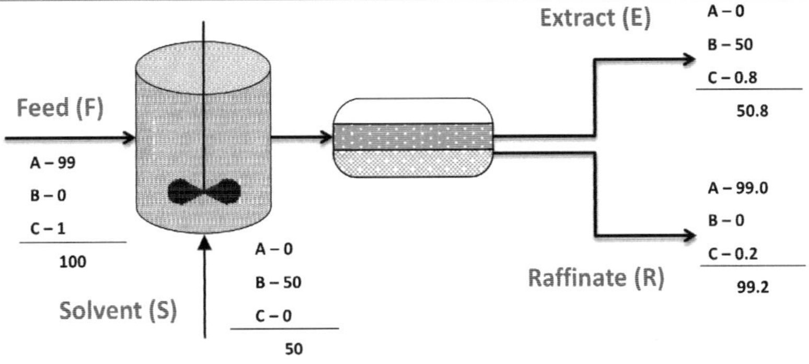

Fraction Unextracted ➤ $U = \frac{\text{Solute in Raffinate}}{\text{Solute in Feed}} = \frac{0.2}{1.0} = 0.2$

Distribution Coefficient ➤ $M = \frac{\text{Conc. Solute in Extract}}{\text{Conc. Solute in Raffinate}} = \frac{0.8/50}{0.2/99} = 7.92$

Extraction Factor ➤ $E = (S/F)(M) = (50/99)(7.92) = 4.0$

2

www.modularprocess.com
Telephone: (201) 267-8670

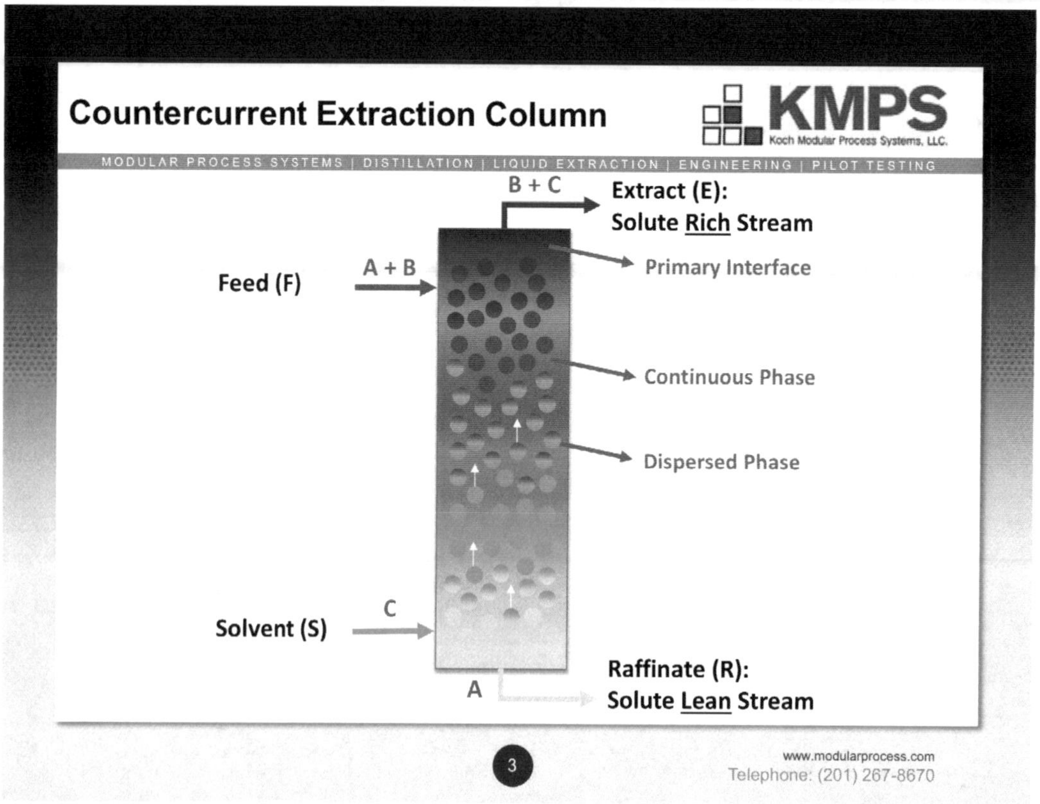
Countercurrent Extraction Column
KMPS
Koch Modular Process Systems, LLC.
MODULAR PROCESS SYSTEMS | DISTILLATION | LIQUID EXTRACTION | ENGINEERING | PILOT TESTING
B + C
Extract (E):
Solute Rich Stream
Primary Interface
Feed (F)
A + B
Continuous Phase
Dispersed Phase
Solvent (S)
C
Raffinate (R):
Solute Lean Stream
A
3
www.modularprocess.com
Telephone: (201) 267-8670

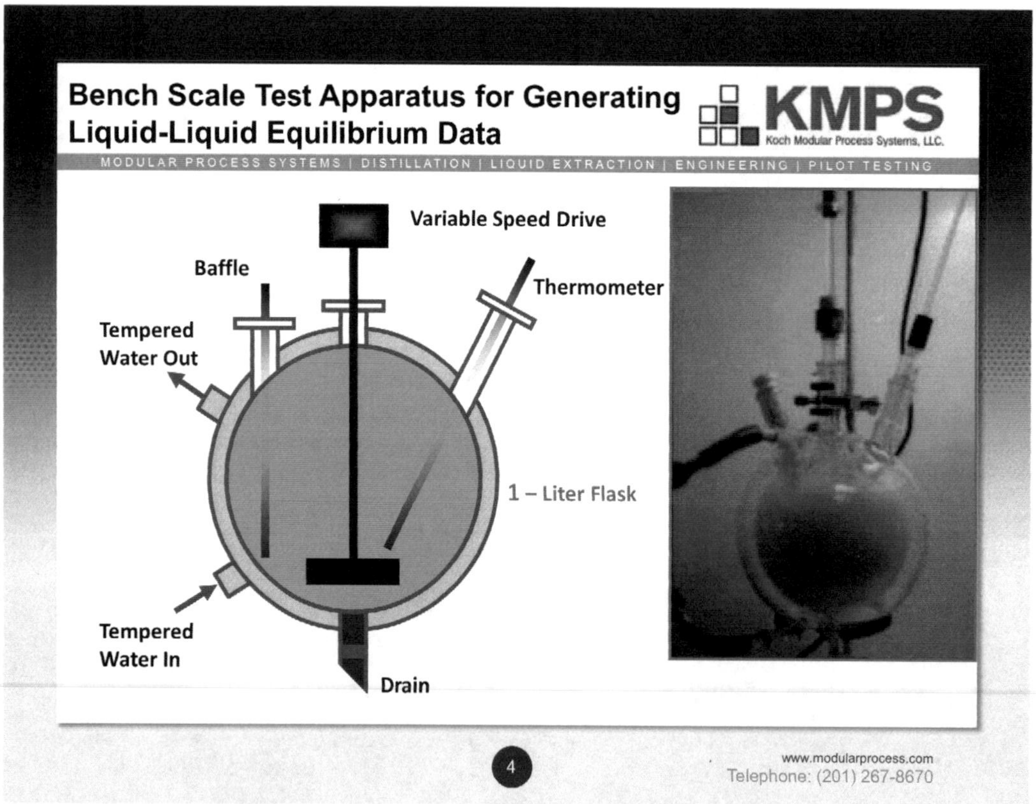
Bench Scale Test Apparatus for Generating Liquid-Liquid Equilibrium Data
KMPS
Koch Modular Process Systems, LLC.
MODULAR PROCESS SYSTEMS | DISTILLATION | LIQUID EXTRACTION | ENGINEERING | PILOT TESTING
Variable Speed Drive
Baffle
Thermometer
Tempered
Water Out
1 – Liter Flask
Tempered
Water In
Drain
4
www.modularprocess.com
Telephone: (201) 267-8670

LLE Equilibrium and Operating Lines

KMPS Koch Modular Process Systems, LLC.

MODULAR PROCESS SYSTEMS | DISTILLATION | LIQUID EXTRACTION | ENGINEERING | PILOT TESTING

Solute Free Basis

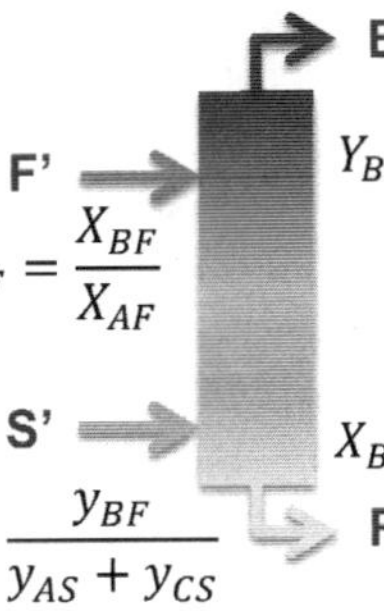

$$Y_{BE} = \frac{y_{BE}}{y_{AR} + y_{CE}}$$

$$X_{BF} = \frac{X_{BF}}{X_{AF}}$$

$$X_{BR} = \frac{x_{BR}}{x_{AR} + x_{CR}}$$

$$Y_{BS} = \frac{y_{BF}}{y_{AS} + y_{CS}}$$

$$F' = F(x_{AR})$$
$$S' = S(y_{AS}y_{CS})$$
$$E' = E(y_{AE}y_{CE})$$
$$R' = R(x_{AR}x_{CR})$$

Graphical Solution

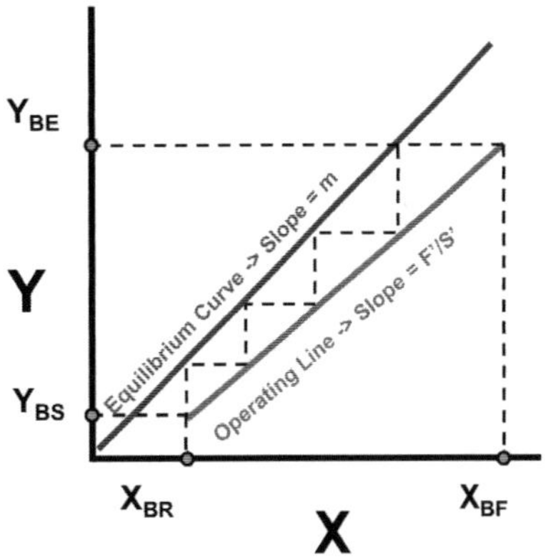

$$m = \frac{Y_{B^*}}{X_{B^*}} \left(\begin{array}{c}\text{Distribution Coefficient} \\ \text{on Solute Free Basis}\end{array}\right)$$

5

www.modularprocess.com
Telephone: (201) 267-8670

Typical LLE Equilibrium Curve

KMPS Koch Modular Process Systems, LLC.

MODULAR PROCESS SYSTEMS | DISTILLATION | LIQUID EXTRACTION | ENGINEERING | PILOT TESTING

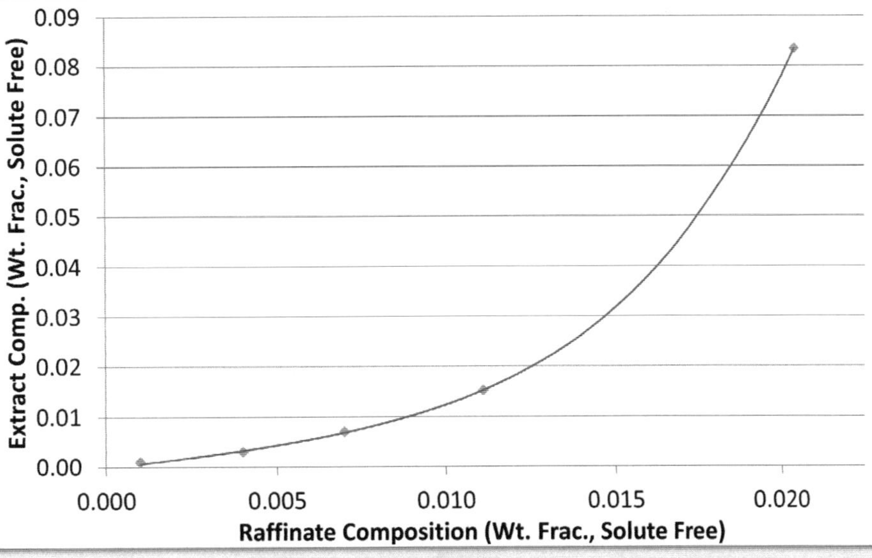

6

www.modularprocess.com
Telephone: (201) 267-8670

Graphical Determination of Theoretical Stages

98% Solute Extraction, S/F = 1.0 mass basis

KMPS Koch Modular Process Systems, LLC.

MODULAR PROCESS SYSTEMS | DISTILLATION | LIQUID EXTRACTION | ENGINEERING | PILOT TESTING

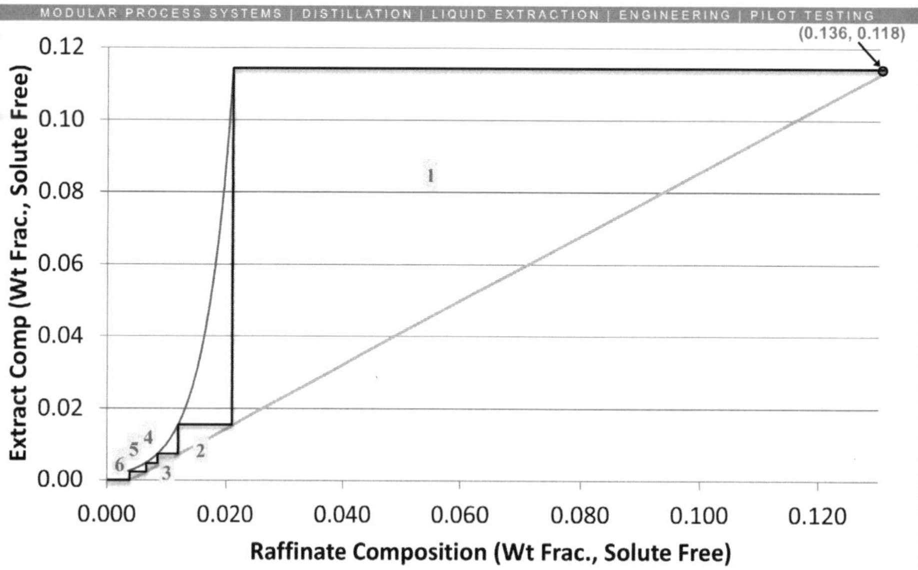

7

www.modularprocess.com
Telephone: (201) 267-8670

Kremser Equation

KMPS Koch Modular Process Systems, LLC.

MODULAR PROCESS SYSTEMS | DISTILLATION | LIQUID EXTRACTION | ENGINEERING | PILOT TESTING

$$n = \frac{\mathrm{LOG}\left[\left(\frac{x_f - {y_s}/{m}}{x_n - {y_s}/{m}}\right)\left(1 - \frac{1}{E}\right) + \frac{1}{E}\right]}{\mathrm{LOG}\,E}$$

Where:
- n = Number of theoretical stages required
- x_f = Conc. of solute in feed on solute free basis
- x_n = Conc. of solute in raffinate on solute free basis
- y_s = Conc. of solute in solvent on solute free basis
- m = Distribution coefficient
- E = Extraction factor = (m)(S/F)

8

www.modularprocess.com
Telephone: (201) 267-8670

Liquid-Liquid Extraction Design

KMPS Koch Modular Process Systems, LLC.

MODULAR PROCESS SYSTEMS | DISTILLATION | LIQUID EXTRACTION | ENGINEERING | PILOT TESTING

Theoretical design is difficult and can be inaccurate due to:

Complex processes occurring in an extractor

- Coalescing / Wetting Characteristics
- Interfacial and Drop Turbulence Effects
- Axial and Radial Mixing
- Tendency to Emulsify
- Phase Ratio Variability and Density Gradients
- Effects of Impurities or Solids
- Mass Transfer Rate

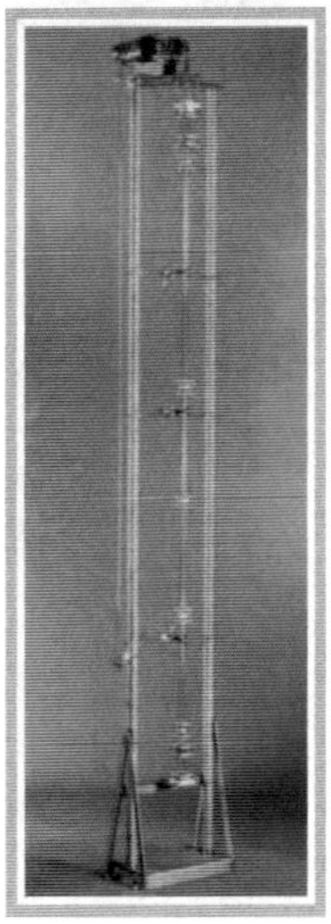

9

www.modularprocess.com
Telephone: (201) 267-8670

Liquid-Liquid Extraction Scale-Up

KMPS Koch Modular Process Systems, LLC.

MODULAR PROCESS SYSTEMS | DISTILLATION | LIQUID EXTRACTION | ENGINEERING | PILOT TESTING

Best method of design:

Pilot testing followed by empirical scale-up

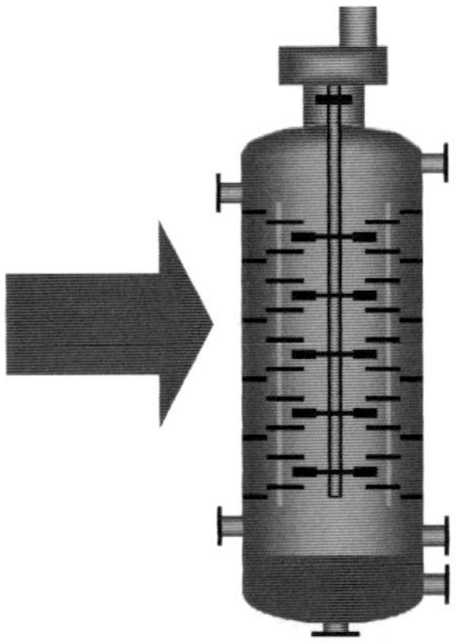

10

www.modularprocess.com
Telephone: (201) 267-8670

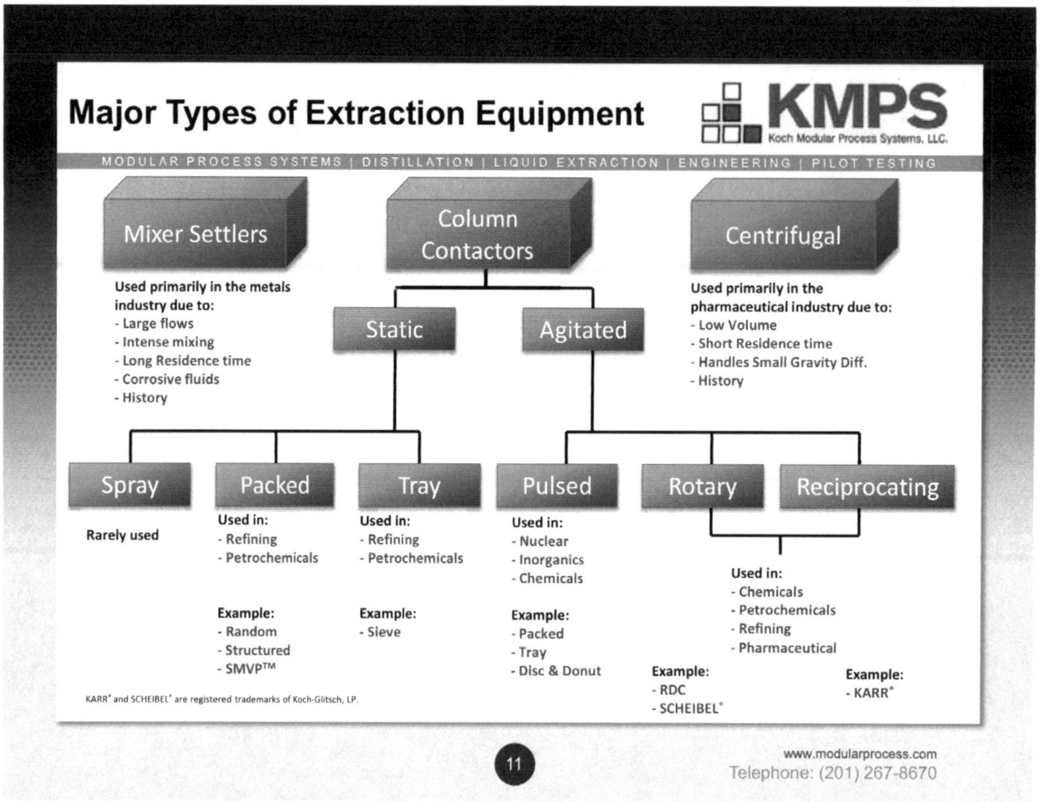

Major Types of Extraction Equipment
KMPS
Koch Modular Process Systems, LLC.
MODULAR PROCESS SYSTEMS | DISTILLATION | LIQUID EXTRACTION | ENGINEERING | PILOT TESTING
Mixer Settlers
Column Contactors
Centrifugal
Used primarily in the metals industry due to:
- Large flows
- Intense mixing
- Long Residence time
- Corrosive fluids
- History
Static
Agitated
Used primarily in the pharmaceutical industry due to:
- Low Volume
- Short Residence time
- Handles Small Gravity Diff.
- History
Spray
Packed
Tray
Pulsed
Rotary
Reciprocating
Rarely used
Used in:
- Refining
- Petrochemicals
Used in:
- Refining
- Petrochemicals
Used in:
- Nuclear
- Inorganics
- Chemicals
Used in:
- Chemicals
- Petrochemicals
- Refining
- Pharmaceutical
Example:
- Random
- Structured
- SMVP™
Example:
- Sieve
Example:
- Packed
- Tray
- Disc & Donut
Example:
- RDC
- SCHEIBEL®
Example:
- KARR®
KARR® and SCHEIBEL® are registered trademarks of Koch-Glitsch, LP.
11
www.modularprocess.com
Telephone: (201) 267-8670

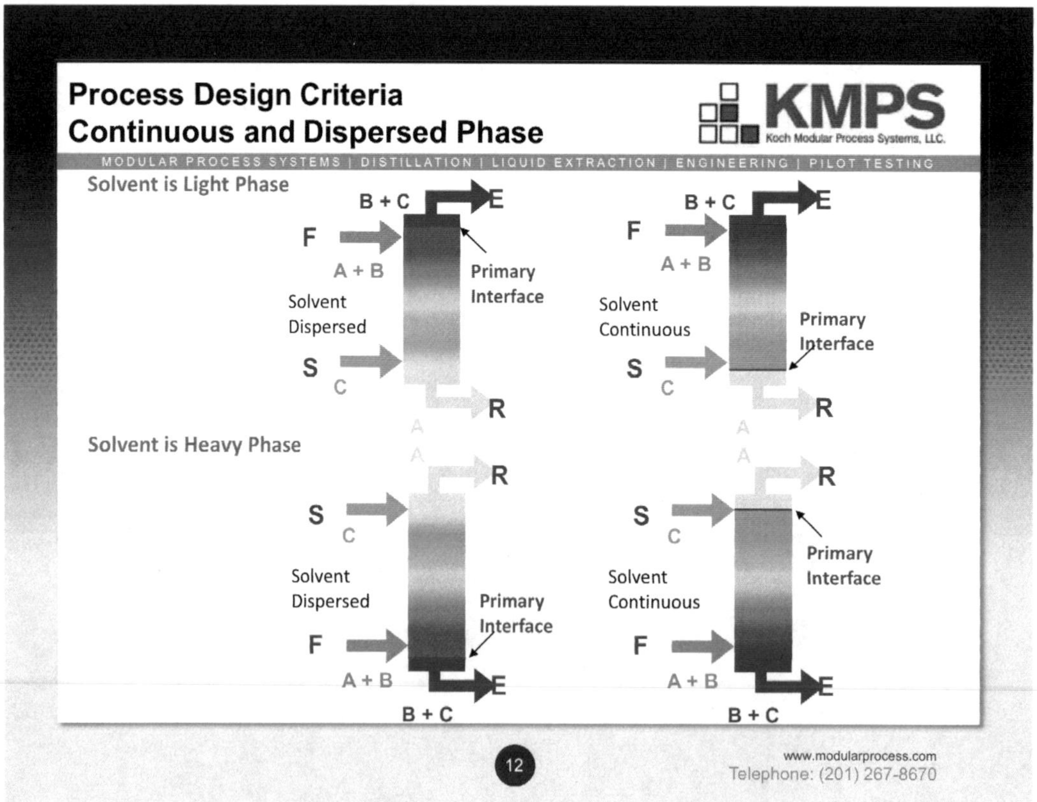

Process Design Criteria
Continuous and Dispersed Phase
KMPS
Koch Modular Process Systems, LLC.
MODULAR PROCESS SYSTEMS | DISTILLATION | LIQUID EXTRACTION | ENGINEERING | PILOT TESTING
Solvent is Light Phase
B + C
E
F
A + B
Primary Interface
Solvent Dispersed
S
C
R
A
B + C
E
F
A + B
Solvent Continuous
Primary Interface
S
C
R
A
Solvent is Heavy Phase
A
R
S
C
Solvent Dispersed
Primary Interface
F
A + B
E
B + C
A
R
S
C
Primary Interface
Solvent Continuous
F
A + B
E
B + C
12
www.modularprocess.com
Telephone: (201) 267-8670

Determining the Dispersed Phase

KMPS Koch Modular Process Systems, LLC.

MODULAR PROCESS SYSTEMS | DISTILLATION | LIQUID EXTRACTION | ENGINEERING | PILOT TESTING

Flow Rate (Phase Ratio)

- For Sieve Tray and Packed Columns – disperse the higher flowing phase
- For Agitated Columns – disperse lower flowing phase

Viscosity

- For efficiency – disperse less viscous phase

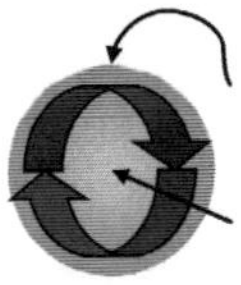

- For capacity – disperse more viscous phase

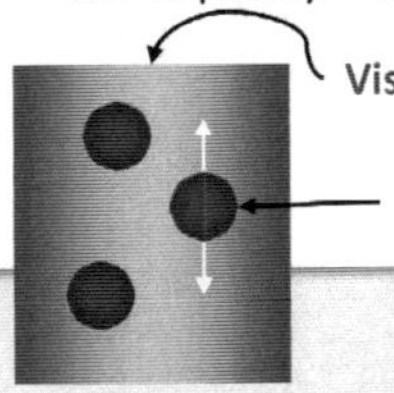

13

www.modularprocess.com
Telephone: (201) 267-8670

Determining the Dispersed Phase

KMPS Koch Modular Process Systems, LLC.

MODULAR PROCESS SYSTEMS | DISTILLATION | LIQUID EXTRACTION | ENGINEERING | PILOT TESTING

Surface Wetting

- Want the continuous phase to preferentially wet the internals – this minimizes coalescence and therefore maximizes interfacial area

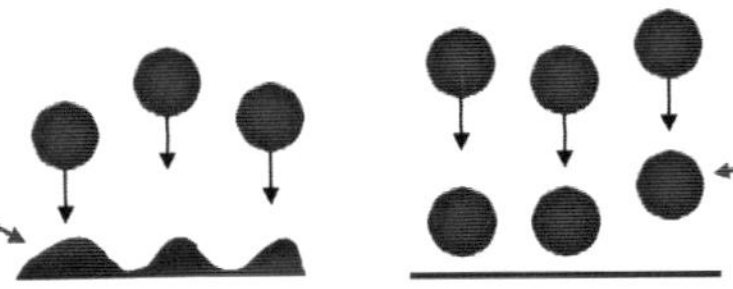

- Rules of Thumb for aqueous-organic systems:
  - Use metal internals when water is continuous phase
  - Use plastic internals (PTFE) when organic is continuous phase

However, this doesn't always work!

14

www.modularprocess.com
Telephone: (201) 267-8670

Determining the Dispersed Phase

KMPS Koch Modular Process Systems, LLC.

MODULAR PROCESS SYSTEMS | DISTILLATION | LIQUID EXTRACTION | ENGINEERING | PILOT TESTING

Marangoni Effect

- Coalescence is enhanced by mass transfer from droplets ⟶ continuous phase

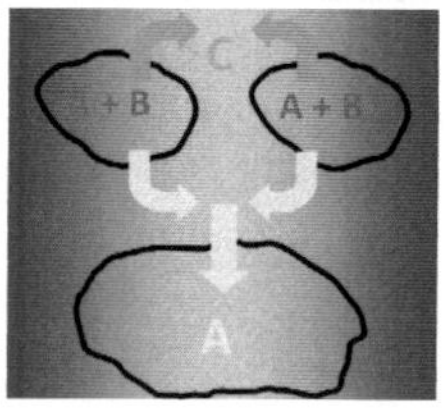

Mass Transfer Direction

Dispersed ⟶ Continuous
(d ⟶ c)

- Droplets tend to coalesce
- Must be counteracted by additional energy

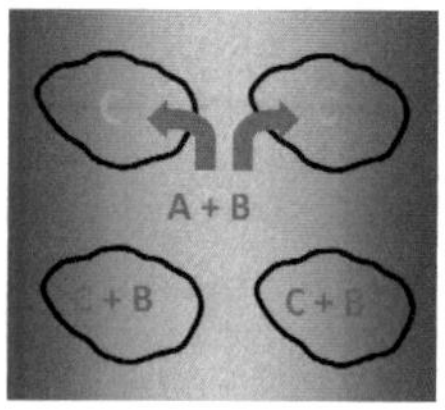

Continuous ⟶ Dispersed
(c ⟶ d)

- Droplets tend to repel each other
- Less energy required to maintain dispersion

15

www.modularprocess.com
Telephone: (201) 267-8670

Interface Behavior

KMPS Koch Modular Process Systems, LLC.

MODULAR PROCESS SYSTEMS | DISTILLATION | LIQUID EXTRACTION | ENGINEERING | PILOT TESTING

Solids build up at the interface = Rag Layer

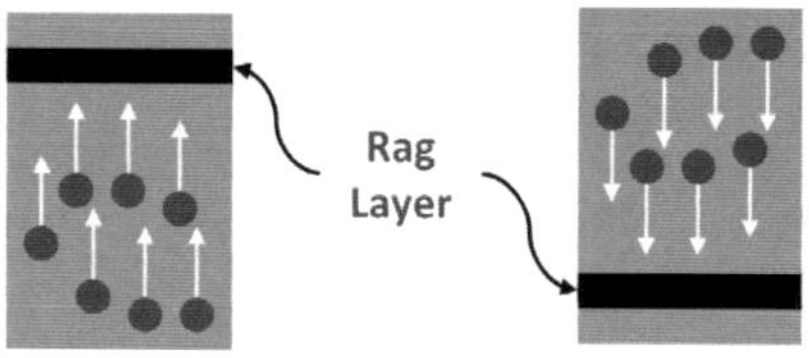

Emulsion band builds up at interface due to slow phase separation

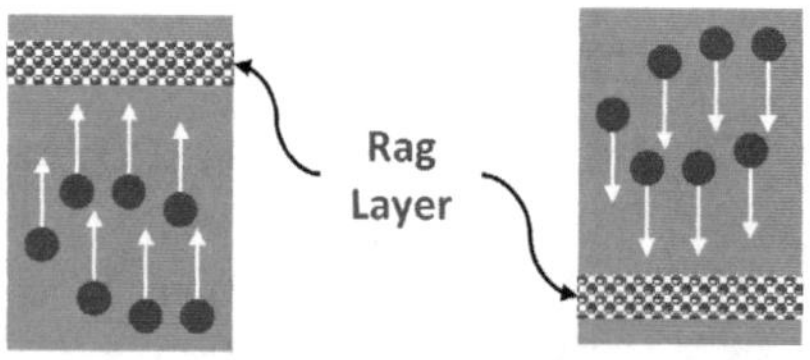

16

www.modularprocess.com
Telephone: (201) 267-8670

Rag Layer or Emulsion Band

KMPS Koch Modular Process Systems, LLC.

MODULAR PROCESS SYSTEMS | DISTILLATION | LIQUID EXTRACTION | ENGINEERING | PILOT TESTING

Corrective Action

1. Reverse Phases
2. For Emulsion Band: slow down capacity or reduce agitation speed
3. For Rag Layer: circulate liquid through an external filter

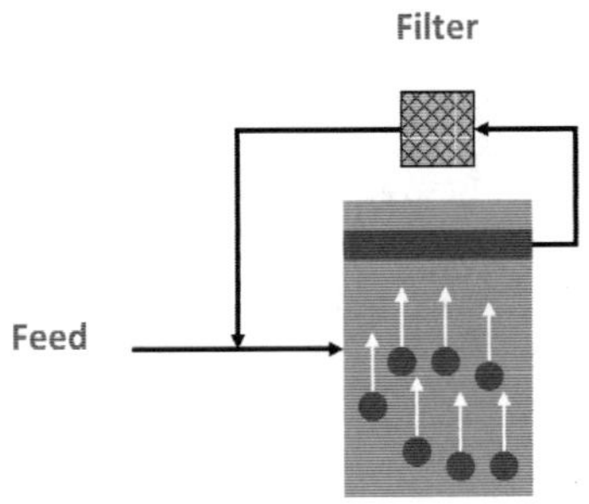

www.modularprocess.com
Telephone: (201) 267-8670

17

Entrainment

KMPS Koch Modular Process Systems, LLC.

MODULAR PROCESS SYSTEMS | DISTILLATION | LIQUID EXTRACTION | ENGINEERING | PILOT TESTING

Entrainment refers to removing a small portion of one phase out of the wrong end of the column i.e. where the other phase exits. Entrainment is controlled by:

1.) Increased settling time inside the column
2.) Coalescer inside the column
3.) Coalescer external to the column

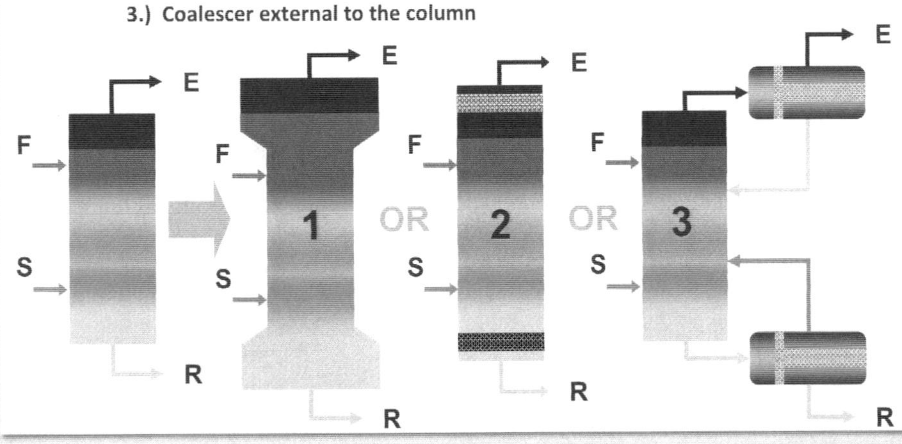

www.modularprocess.com
Telephone: (201) 267-8670

18

Flooding

MODULAR PROCESS SYSTEMS | DISTILLATION | LIQUID EXTRACTION | ENGINEERING | PILOT TESTING

Flooding – the point where the upward or downward flow of the dispersed phase ceases and a second interface is formed in the column

Flooding can be caused by:

- **Increased continuous phase flow rate which increases drag on droplets**

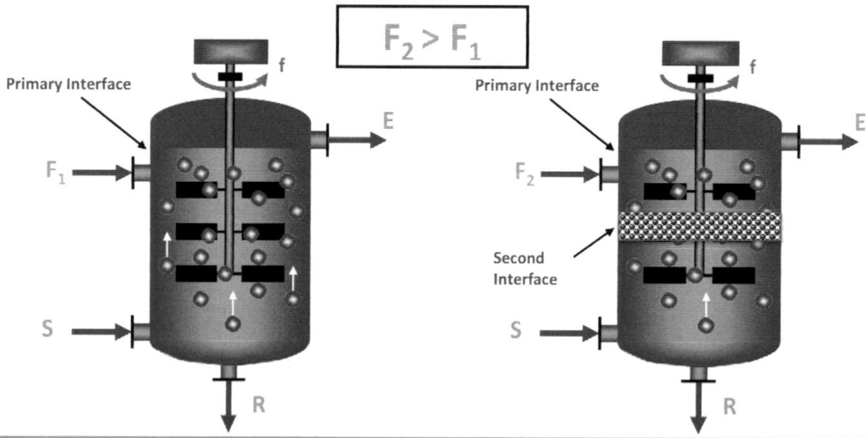

www.modularprocess.com
Telephone: (201) 267-8670

Flooding

MODULAR PROCESS SYSTEMS | DISTILLATION | LIQUID EXTRACTION | ENGINEERING | PILOT TESTING

Flooding can be caused by:

- **Increased agitation speed which forms smaller droplets which cannot overcome flow of the continuous phase**
- **Decreased interfacial tension – forms smaller drops – same effect as increased agitation**

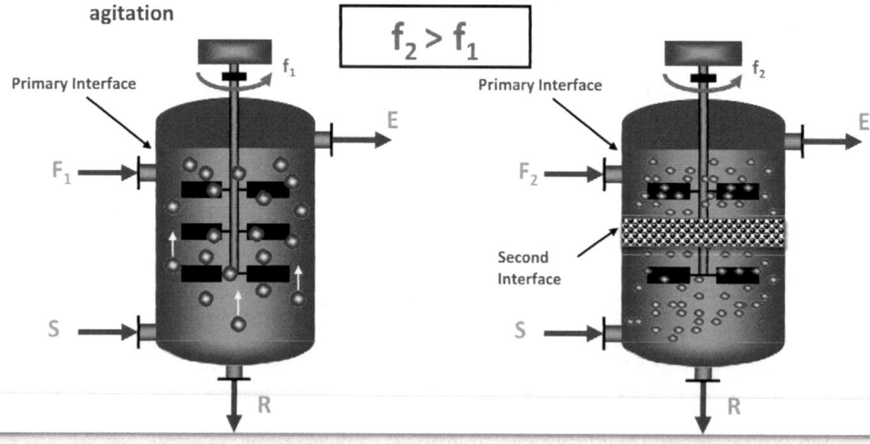

www.modularprocess.com
Telephone: (201) 267-8670

Variables to Consider During Testing

KMPS Koch Modular Process Systems, LLC.

MODULAR PROCESS SYSTEMS | DISTILLATION | LIQUID EXTRACTION | ENGINEERING | PILOT TESTING

- Continuous vs dispersed phase
- Interface behavior – rag layer or emulsion band
- Entrainment – is this an issue and how can it be minimized
- Flooding point – this affects the column capacity (diameter)
- Variable hydraulic behavior throughout the column
- Optimal temperature and pH
- Efficiency vs capacity
  Efficiency vs agitation speed

21

www.modularprocess.com
Telephone: (201) 267-8670

Extractor Flow Patterns

KMPS Koch Modular Process Systems, LLC.

MODULAR PROCESS SYSTEMS | DISTILLATION | LIQUID EXTRACTION | ENGINEERING | PILOT TESTING

Ideal Plug Flow **Actual Flow**

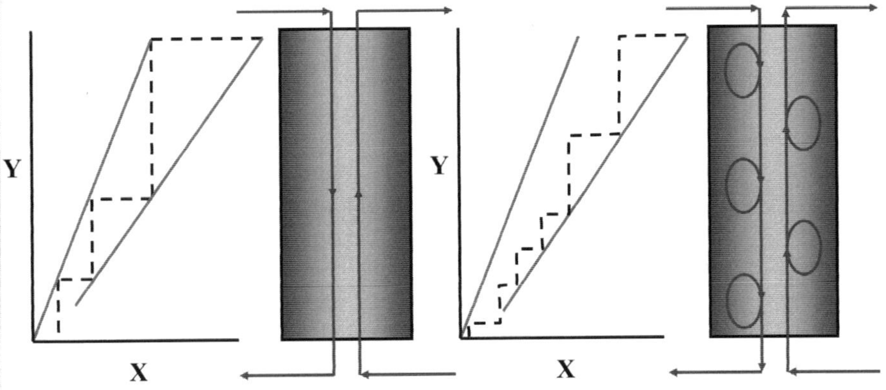

This "axial" or "back" mixing causes concentration gradients that decrease driving force and therefore increase HETS

22

www.modularprocess.com
Telephone: (201) 267-8670

Generalized Scale-up Procedure

KMPS Koch Modular Process Systems, LLC.

MODULAR PROCESS SYSTEMS | DISTILLATION | LIQUID EXTRACTION | ENGINEERING | PILOT TESTING

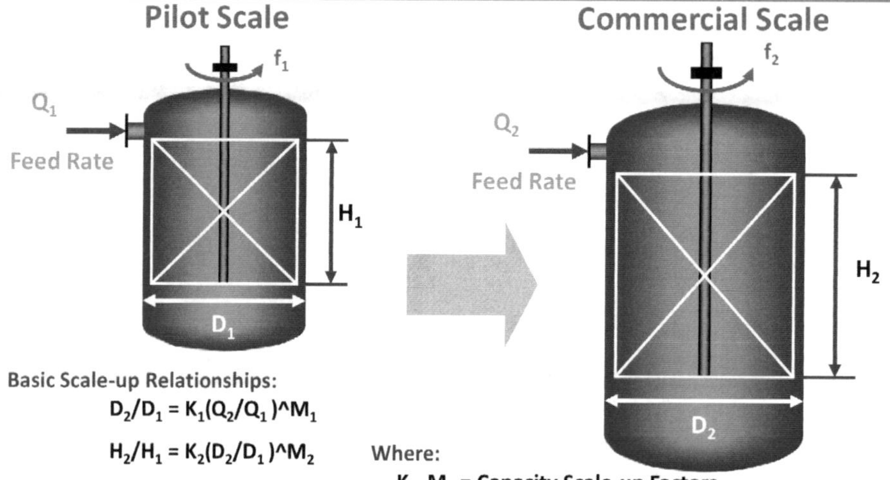

Basic Scale-up Relationships:

$D_2/D_1 = K_1(Q_2/Q_1)$^M_1

$H_2/H_1 = K_2(D_2/D_1)$^M_2

$f_2/f_1 = K_3(D_2/D_1)$^M_3

Where:

K_1, M_1 = Capacity Scale-up Factors

K_2, M_2 = Efficiency Scale-up Factors

K_3, M_3 = Power Scale-up Factors

23

www.modularprocess.com
Telephone: (201) 267-8670

Continuous Extraction Pilot Plant Arrangement

KMPS Koch Modular Process Systems, LLC.

MODULAR PROCESS SYSTEMS | DISTILLATION | LIQUID EXTRACTION | ENGINEERING | PILOT TESTING

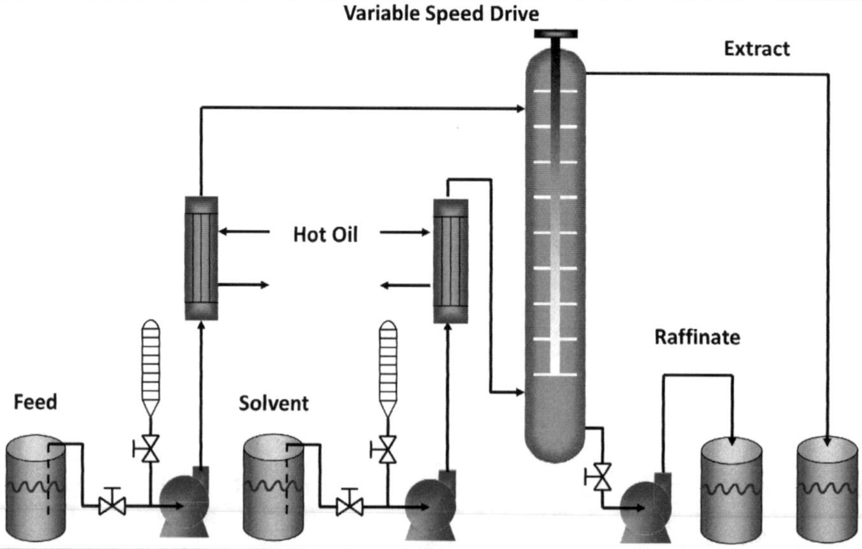

24

www.modularprocess.com
Telephone: (201) 267-8670

Pilot Plant Configuration

KMPS Koch Modular Process Systems, LLC.

MODULAR PROCESS SYSTEMS | DISTILLATION | LIQUID EXTRACTION | ENGINEERING | PILOT TESTING

Determine type of column to be used based on process considerations

Use the same kind of equipment for the production unit

Determine diameter and height of pilot column based on experience

Table 1: Smallest Possible Column for Testing

Type of Column	Diameter	Height
Packed	3″ to 4″	3′ to 6′ per Theoretical Stage (TS)
Tray	≥ 6″	4 to 5 Trays per TS
KARR®	1″	1′ to 3′ per TS
RDC or SCHEIBEL®	3″	3 to 6 Actual Stages per TS (Approx. 3″ to 6″)

KARR® and SCHEIBEL® are registered trademarks of Koch-Glitsch, LP.

25

www.modularprocess.com
Telephone: (201) 267-8670

Pilot Testing – Step by Step Optimization

KMPS Koch Modular Process Systems, LLC.

MODULAR PROCESS SYSTEMS | DISTILLATION | LIQUID EXTRACTION | ENGINEERING | PILOT TESTING

- Select column type and set-up column (multiple inlets)
- Select continuous and dispersed phases
- Select initial operating conditions (capacity – S/F – No. Stages)
- Operate at increasing agitation speed (flooding)
- Based on results, vary:
  1. Continuous phase
  2. S/F vs. No. Stages
  3. Capacity
  4. Other Variables (temp, pH, etc.)

26

www.modularprocess.com
Telephone: (201) 267-8670

Scale-up Example: Recovery of Carboxylic Acids from Fermentation Broth

KMPS Koch Modular Process Systems, LLC.

MODULAR PROCESS SYSTEMS | DISTILLATION | LIQUID EXTRACTION | ENGINEERING | PILOT TESTING

- Broth generated from cellulosic materials – ~5% acids
- LLE Goal: >95% recovery (high purity) and minimal solvent use
- Ethyl acetate solvent – but emulsified easily

Preliminary Data in RDC Columns

- Difficult operation due to emulsification
- < 90% acid recovery
- High S/F ratio – 2.0

KARR® Column Required

KARR® Reciprocating Column

KMPS Koch Modular Process Systems, LLC.

MODULAR PROCESS SYSTEMS | DISTILLATION | LIQUID EXTRACTION | ENGINEERING | PILOT TESTING

Characteristics

- Highest capacity:
  - 30-60 M^3/M^2-hr
  - 750-1,500 gal/ft^2-hr
- High efficiency
- Good turndown capability (4:1)
- Uniform shear mixing
- Best suited for systems that emulsify

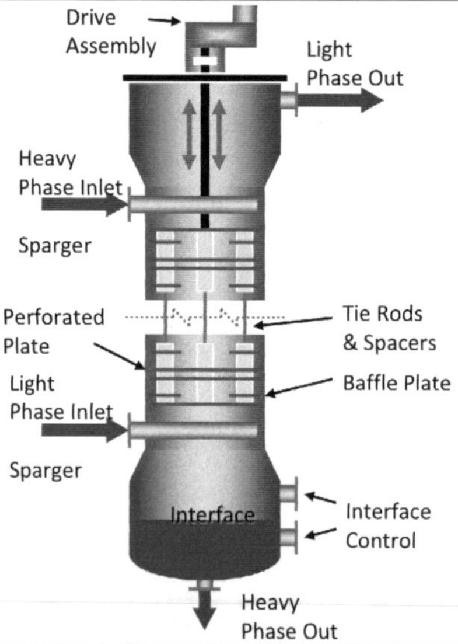

KARR® Column Plate Stack Assembly

KMPS Koch Modular Process Systems, LLC.

MODULAR PROCESS SYSTEMS | DISTILLATION | LIQUID EXTRACTION | ENGINEERING | PILOT TESTING

29

www.modularprocess.com
Telephone: (201) 267-8670

Pilot KARR® Column for Carboxylic Acids 1" diameter x 12' Plate Stack

KMPS Koch Modular Process Systems, LLC.

MODULAR PROCESS SYSTEMS | DISTILLATION | LIQUID EXTRACTION | ENGINEERING | PILOT TESTING

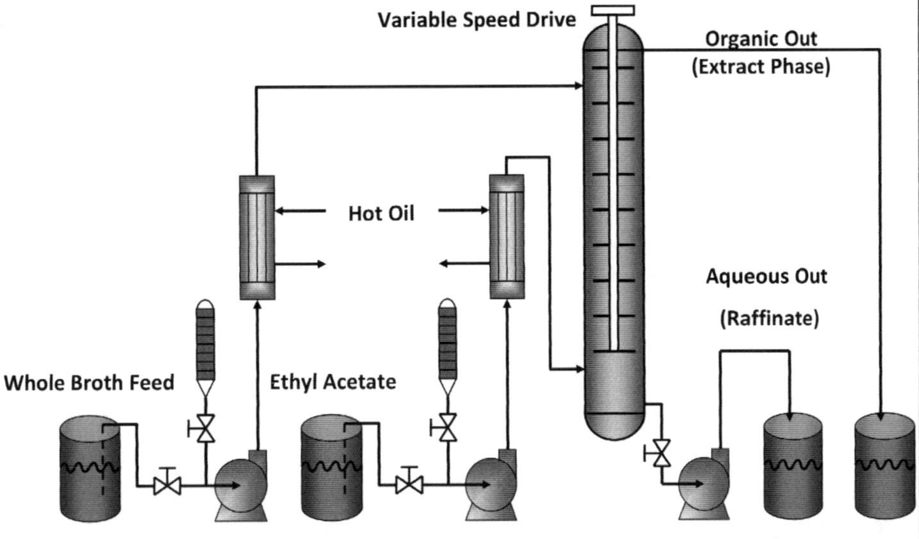

30

www.modularprocess.com
Telephone: (201) 267-8670

Extraction Testing Results

KMPS Koch Modular Process Systems, LLC.

MODULAR PROCESS SYSTEMS | DISTILLATION | LIQUID EXTRACTION | ENGINEERING | PILOT TESTING

- KARR® Column is correct design for this process which failed in RDC column
- S/F ratio improved to 1.5 (client expected 2)
- 96-97% recovery at room temperature and 98-99% recovery at 40-45 C
- Design capacity 650 – 700 GPH/ft^2
- Complete system (LLE + distillations) designed, fabricated, and installed
- LLE testing identified "unknown" component

31

www.modularprocess.com
Telephone: (201) 267-8670

KARR® Column Pilot Plant Scale-up Procedure: Carboxylic Acid Extraction

KMPS Koch Modular Process Systems, LLC.

MODULAR PROCESS SYSTEMS | DISTILLATION | LIQUID EXTRACTION | ENGINEERING | PILOT TESTING

- Select optimal run from test results
  - Runs 7 and 11:
    - Solvent to Feed Ratio (S/F) = 1.5
    - Specific Throughput (Q) = 650 GPH/FT^2
- Production column design
  - Diameter – direct scale-up based on specific throughput
  - Broth Feed = 4974 lb/hr, Ethyl Acetate = 7461 lb/hr
  - Area = (Feed rate + Solvent Rate) / 650 GPH/FT^2
  - Height: $H_{COMM} = f\,(H)_{PILOT}$
  - Agitation Speed: $SPM_{COMM} = f\,(SPM)_{PILOT}$

32

www.modularprocess.com
Telephone: (201) 267-8670

KARR® Column Pilot Plant Scale-up Procedure: Carboxylic Acid Extraction

KMPS Koch Modular Process Systems, LLC.

MODULAR PROCESS SYSTEMS | DISTILLATION | LIQUID EXTRACTION | ENGINEERING | PILOT TESTING

- $H_{COMM} = (D_{COMM} / D_{PILOT})^{0.38} \times H_{PILOT}$
- $H_{COMM} = (22/1)^{0.38} \times (12 \text{ feet}) = 38 \text{ feet}$
- $SPM_{COMM} = (D_{PILOT} / D_{COMM})^{0.14} \times SPM_{PILOT}$
- $SPM_{COMM} = (1/22)^{0.14} \times (50 \text{ SPM}) = 32 \text{ SPM}$
- Where:
  - H_{COMM} = Height Commercial Column
  - H_{PILOT} = Height Pilot Column
  - D_{COMM} = Diameter Commercial Column
  - D_{PILOT} = Diameter Pilot Column
  - SPM_{COMM} = Commercial Strokes Per Minute
  - SPM_{PILOT} = Pilot Strokes Per Minute

33

www.modularprocess.com
Telephone: (201) 267-8670

KARR® Column Design Carboxylic Acid Extraction from Broth

KMPS Koch Modular Process Systems, LLC.

MODULAR PROCESS SYSTEMS | DISTILLATION | LIQUID EXTRACTION | ENGINEERING | PILOT TESTING

- **Diameter = 22" (D_1)**
- **Expanded Ends Diameter = 44" (D_2)**
- **Plate Stack = 38'-0" (A)**
- **Overall Height = 50'-6" (B)**

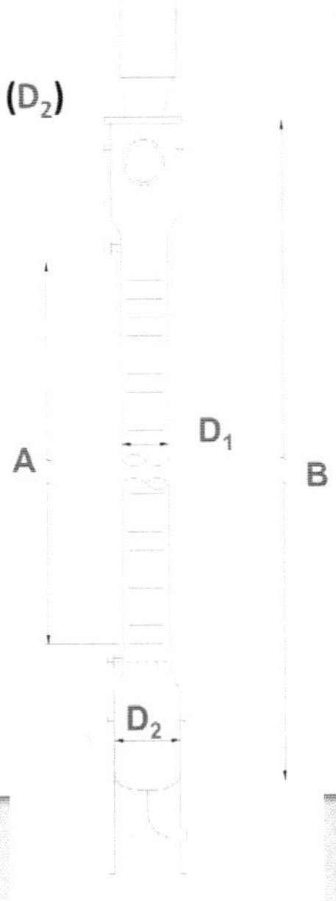

www.modularprocess.com
Telephone: (201) 267-8670

SCHEIBEL® Column

KMPS Koch Modular Process Systems, LLC.

MODULAR PROCESS SYSTEMS | DISTILLATION | LIQUID EXTRACTION | ENGINEERING | PILOT TESTING

Characteristics

- Reasonable capacity:
  15-25 M^3/M^2-hr
  350-600 gal/ft^2-hr
- High efficiency due to internal baffling
- Good turndown capability (4:1) and high flexibility
- Best suited when many stages are required
- Not recommended for highly fouling systems or systems that tend to emulsify

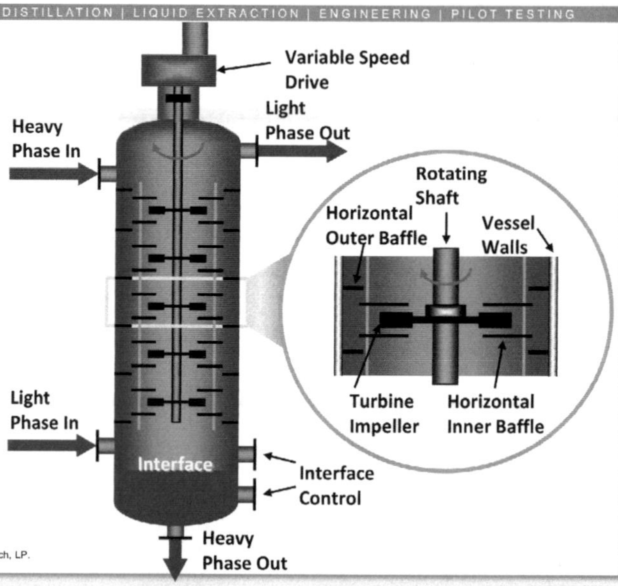

KARR® and SCHEIBEL® are registered trademarks of Koch-Glitsch, LP.

35

www.modularprocess.com
Telephone: (201) 267-8670

SCHEIBEL® Column Pilot Plant Setup
Nitrated Organics Extraction

KMPS Koch Modular Process Systems, LLC.

MODULAR PROCESS SYSTEMS | DISTILLATION | LIQUID EXTRACTION | ENGINEERING | PILOT TESTING

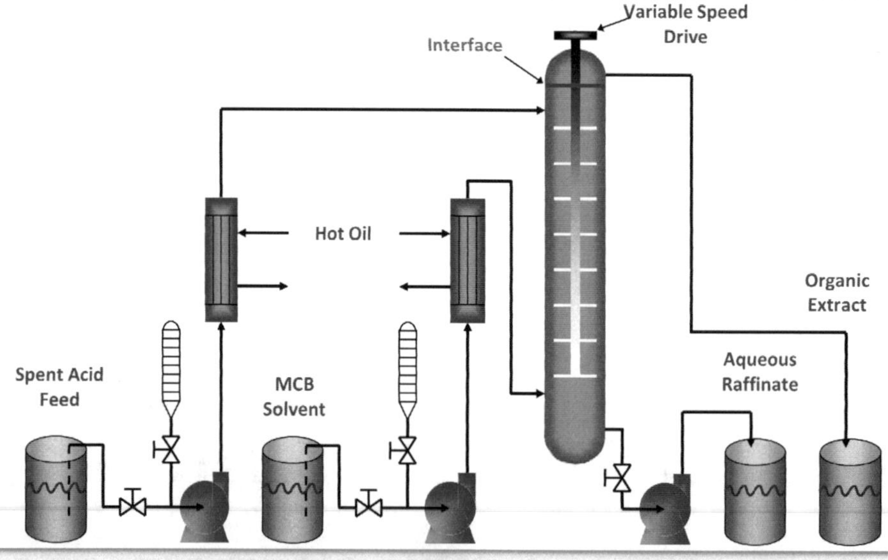

KARR® and SCHEIBEL® are registered trademarks of Koch-Glitsch, LP.

36

www.modularprocess.com
Telephone: (201) 267-8670

SCHEIBEL® Column Scale-up Procedure
Nitrated Organics Extraction

KMPS Koch Modular Process Systems, LLC.

MODULAR PROCESS SYSTEMS | DISTILLATION | LIQUID EXTRACTION | ENGINEERING | PILOT TESTING

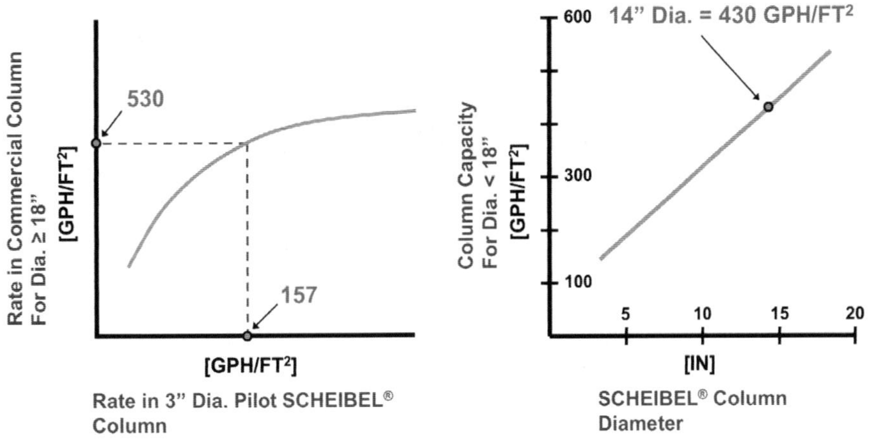

KARR® and SCHEIBEL® are registered trademarks of Koch-Glitsch, LP.

37

www.modularprocess.com
Telephone: (201) 267-8670

SCHEIBEL® Column
Nitrated Organics Extraction

KMPS Koch Modular Process Systems, LLC.

MODULAR PROCESS SYSTEMS | DISTILLATION | LIQUID EXTRACTION | ENGINEERING | PILOT TESTING

- Diameter = 14" (D_1)
- Expanded Head Diameter = 20" (D_2)
- Bed Height = 9'-6" (A)
- Overall Height = 16'-4" (B)

38

www.modularprocess.com
Telephone: (201) 267-8670

Typical Extraction System

KMPS Koch Modular Process Systems, LLC.

MODULAR PROCESS SYSTEMS | DISTILLATION | LIQUID EXTRACTION | ENGINEERING | PILOT TESTING

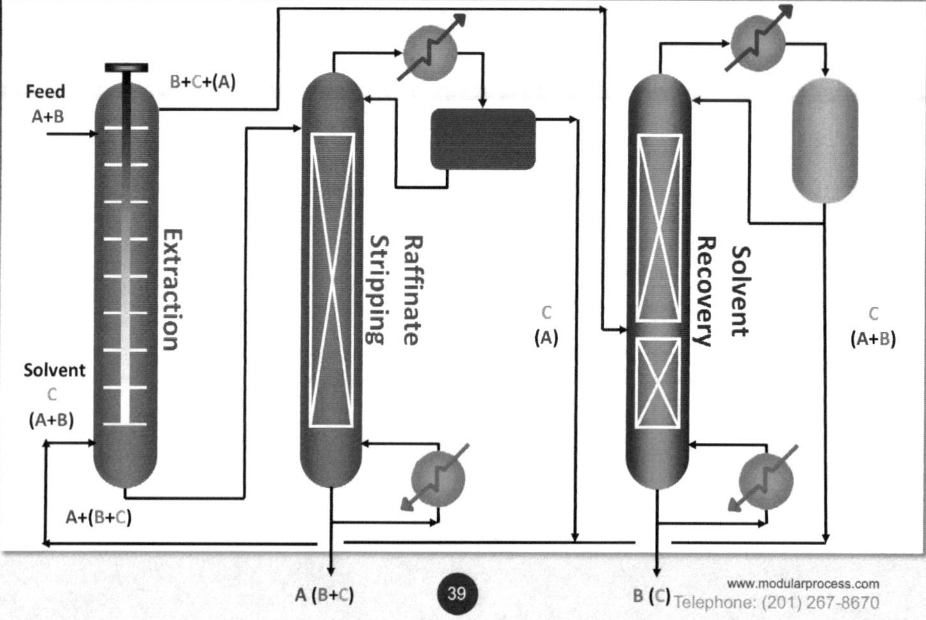

39

www.modularprocess.com
Telephone: (201) 267-8670

Thank You!

KMPS Koch Modular Process Systems, LLC.

MODULAR PROCESS SYSTEMS | DISTILLATION | LIQUID EXTRACTION | ENGINEERING | PILOT TESTING

Questions?

40

www.modularprocess.com
Telephone: (201) 267-8670

Scale-up Systems

The Fast Track for Chemical Process Development

Brains + Experiments + Models = Success

Peter Clark, Scale-up Systems
AIChE Process Development Symposium
Jun 2014

"We've got a few problems going from lab scale to full scale commercial"

DynoChem

Compatible with Windows 7

Microsoft Partner
Gold Independent Software Vendor (ISV)

DynoChem

Models

FDA U.S. Food and Drug Administration
Protecting and Promoting Public Health
www.fda.gov

Introduction

What is a Model?
A model is a representation of an underlying physical/chemical phenomena

Advantages of Models:

- Enhanced process understanding
  - Allows decision making
- Reduction of number of experiments
- Improvement of productivity and product quality

3

Role of Models in the QbD Paradigm: Regulatory Perspective, S. Chatterjee, FDA, AAPS Annual Meeting, October 2011

DynoChem — *Mechanistic models*

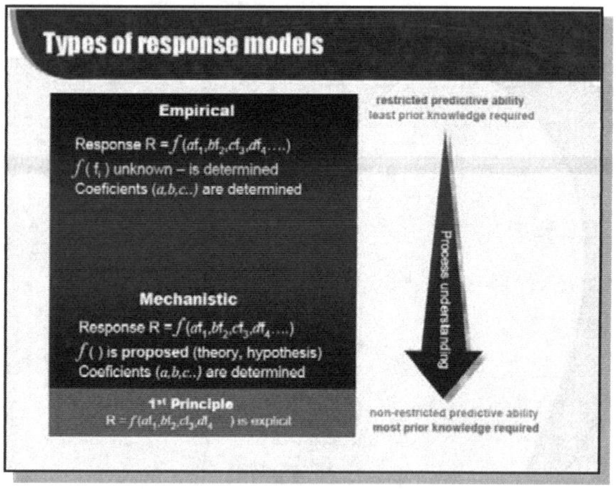

GSK: "Roles of Mechanistic and Empirical Modelling in Achieving QbD" (2009)

DynoChem — *Mechanistic models*

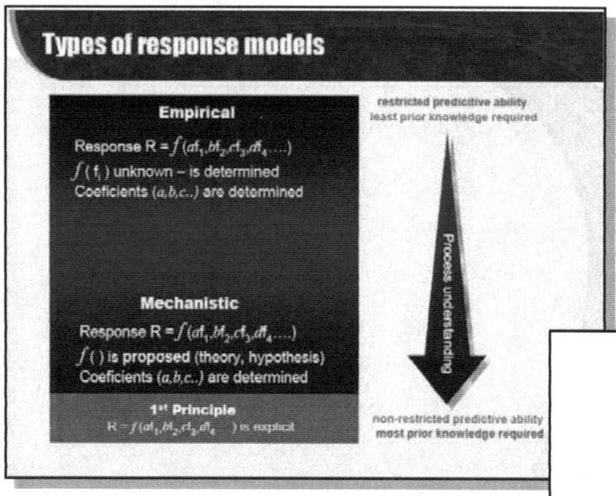

GSK: "Roles of Mechanistic and Empirical Modelling in Achieving QbD" (2009)

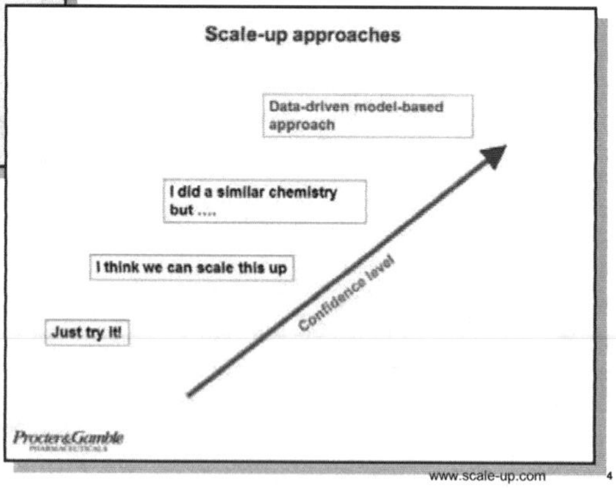

P&G : "A Case Study of Data-Driven and Model-Based Scale-up Approach for API Process Development" (2007)

DynoChem — Case studies

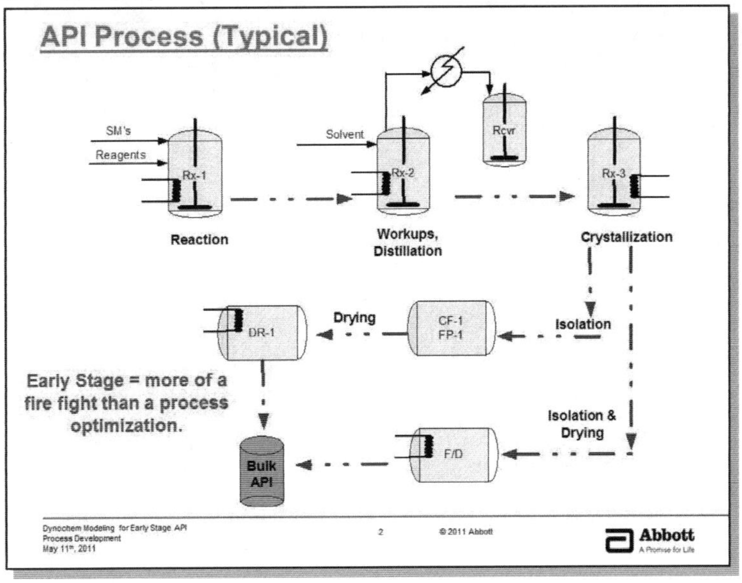

DynoChem Modeling for API Process Development: Early Stage Linking of Laboratory to At-Scale Performance, J. Marek, et. al., Abbott, DynoChem User Meeting 2011

DynoChem — Case studies

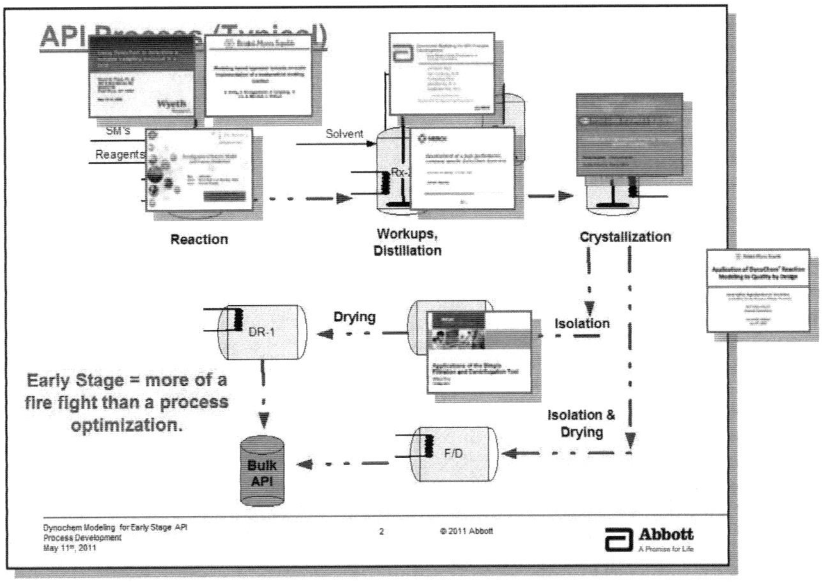

DynoChem Case studies, over the lifecycle

U.S. Food and Drug Administration
Protecting and Promoting Public Health
www.fda.gov

Examples of Models in the QbD Paradigm

Product profile → CQAs → Risk assessment → Design space → Control strategy → Lifecycle Management

Product Design
- Models to optimize formulation
- IV-IVC models
- Reaction kinetic models

Process Design
- Scale up models to scale up parameters from pilot to commercial scale
- Mechanistic and/or empirical models to define design space
- Chemometric models
- Multivariate models to support RTRT (e.g. surrogate models for dissolution)

Lifecycle Management
- MSPC model for continual improvement

19

Role of Models in the QbD Paradigm: Regulatory Perspective, S. Chatterjee, FDA, AAPS Annual Meeting, October 2011

Bristol-Myers Squibb

Modeling based approach towards on-scale implementation of a methanethiol emitting reaction

B. Remy, S. Brueggemeier, O. Lyngberg, D. Lin, A. Marchut, L. Hobson

1

Modeling based approach towards on-scale implementation of a methanethiol emitting reaction, B Remy et. al., BMS, DynoChem User Meeting 2007

DynoChem — BMS: We don't do stinking scale-ups

Scale-up Challenge: Methanethiol By-product Control

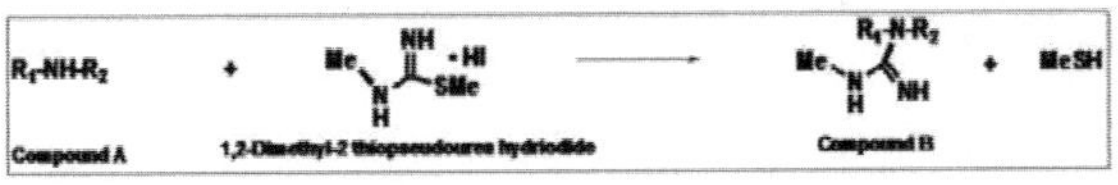

Challenge : Scale-up to 45 kg using standard pilot plant equipment

- Chemistry previously demonstrated at 4.5 kg scale
- Methanethiol gas is generated as by-product
  - Stro
- Rigorou
- Aggres
- Limited

Pilot plant goals:
- *Hit aggressive timeline*
- *Ensure permit compliance*
- *Use single-stage scrubber*
- *Reduce peak off-gas rate*
- *Avoid long reaction time*

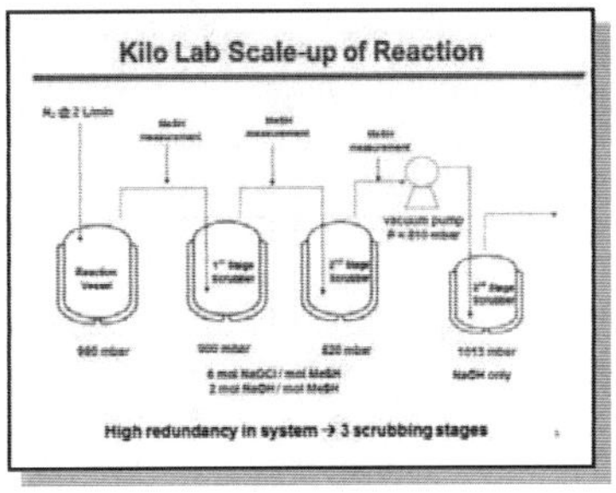

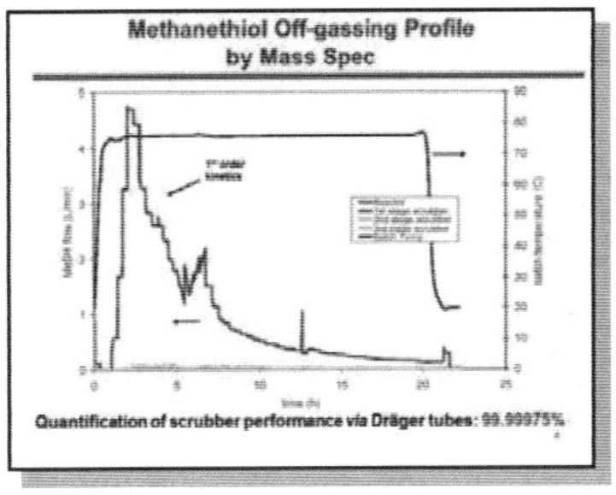

Modeling based approach towards on-scale implementation of a methanethiol emitting reaction, B Remy et. al., BMS, DynoChem User Meeting 2007

DynoChem — Start by drawing a Process Scheme

Modeling of Guanylation Reaction & Methanethiol Off-gassing

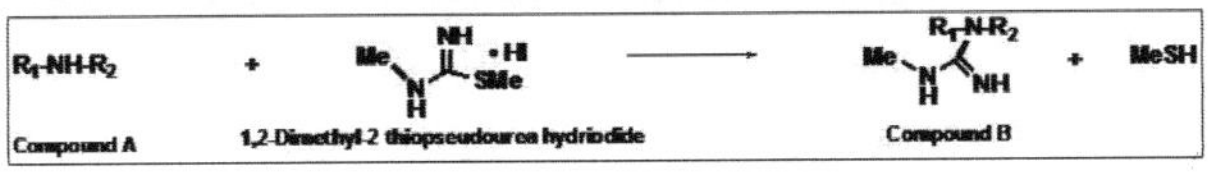

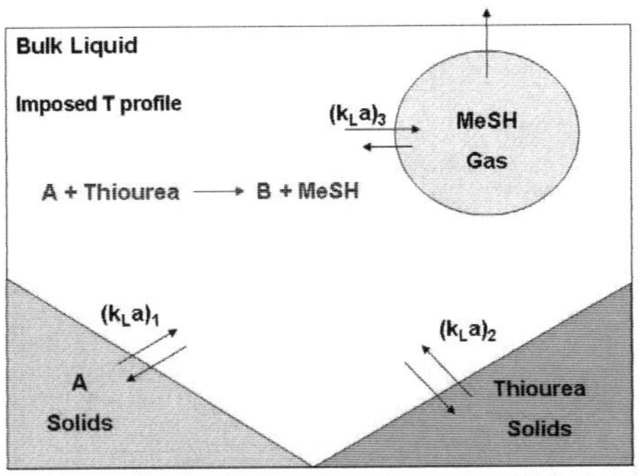

- *Share knowledge and insights about process behavior*
- *Identify scale and equipment dependence*
- *Map out experiments and calculations*

DynoChem *Process schemes help take off blindfolds*

Kidney International (2002) 62, 1524-1538. (Cartoon originally copyrighted by the authors; G. Renee Guzlas, artist).

DynoChem *Build model and fit kinetics*

Modeling of Guanylation Reaction & Methanethiol Off-gassing

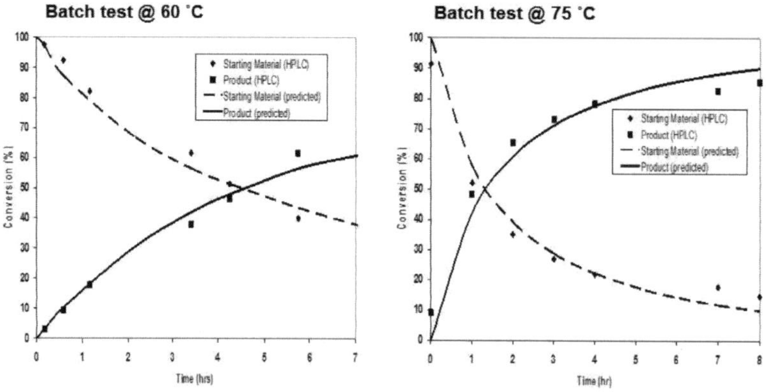

	Final Value	Units	S.E.	C.I (%)	t-Statistic	t crit
k	2.91E-04	L/mol s	1.81E-5	12.15	16.5	2.04
Ea	89.5	kJ/mol	4.1	10.9	18.3	2.04

10

DynoChem — *Choose operating policy*

Simulation of 45 kg Batch Off-gas Rate
Kilo Lab Conditions

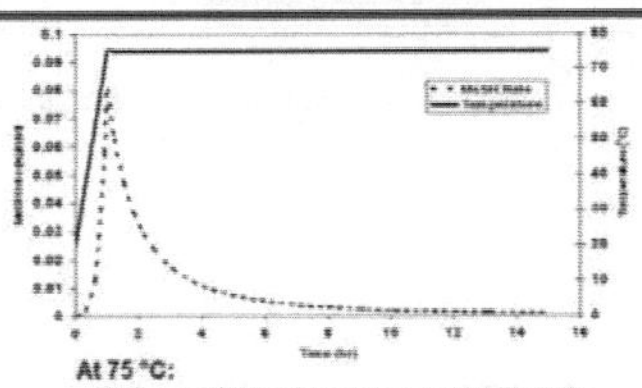

At 75 °C:
- Maximum MeSH off-gas flow rate 0.08 kg/min
- 95% conversion in 13 hr

At 70 °C:
- Maximum MeSH off-gas flow rate 0.04 kg/min
- 95% conversion in 22 hr

Strategy for Reducing Off-gas Rate
Ramp Temperature Profile

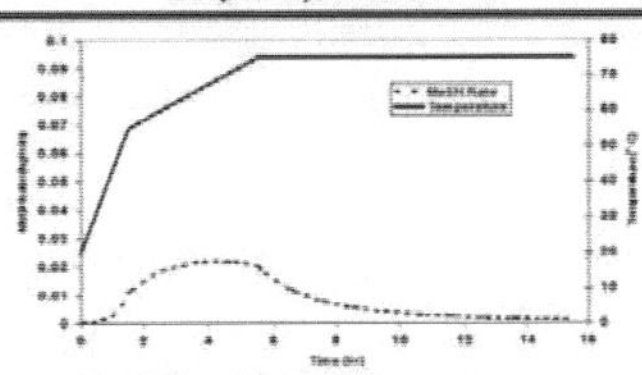

Temperature Ramp 5 °C/hr:
- Maximum MeSH off-gas flow rate 0.02 kg/min
  - 62% reduction vs. Kilo Lab conditions
- 95% conversion in 15 hr
  - 15% increase in reaction time

Strategy for Reducing Off-gas Rate
3 Step Temperature Profile

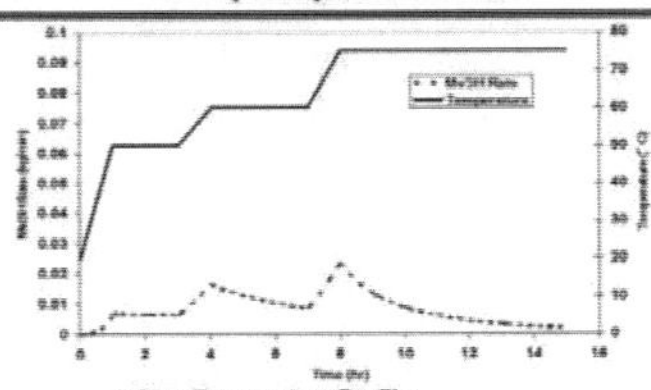

3 Step Temperature Profile:
- Maximum MeSH off-gas flow rate 0.03 kg/min
  - 60% reduction vs. Kilo Lab conditions
- 95% conversion in 16 hr
  - 23% increase in reaction time

Time Required to Reduce MeSH Off-gassing
Reaction Emergency Shutdown

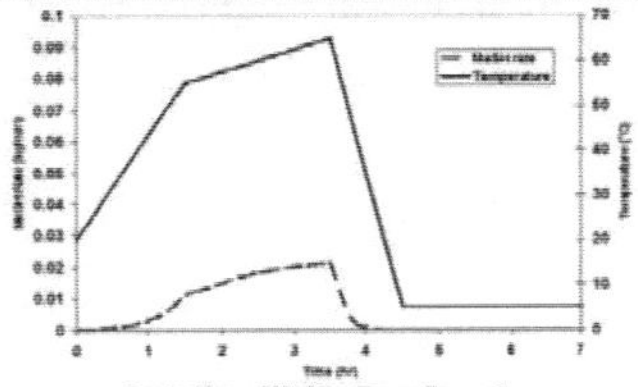

Reduction of MeSH off-gas flow rate:
- Batch cooling at 1 °C/min:
  - 70% in 10 min
  - 95% in 30 min

DynoChem — *Verify model*

Model Prediction vs. Experimental Result
25 g Temperature Step Run

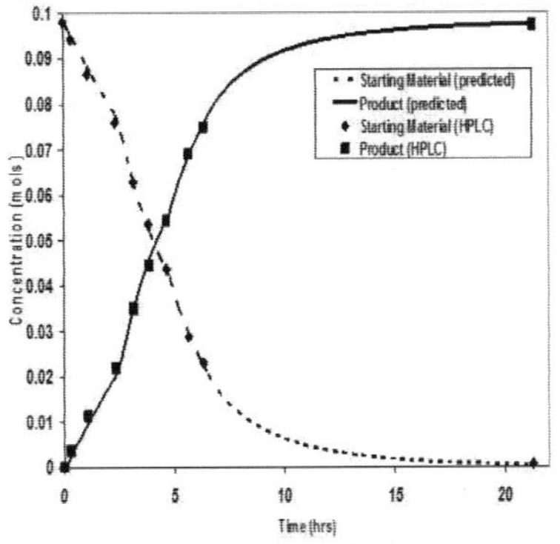

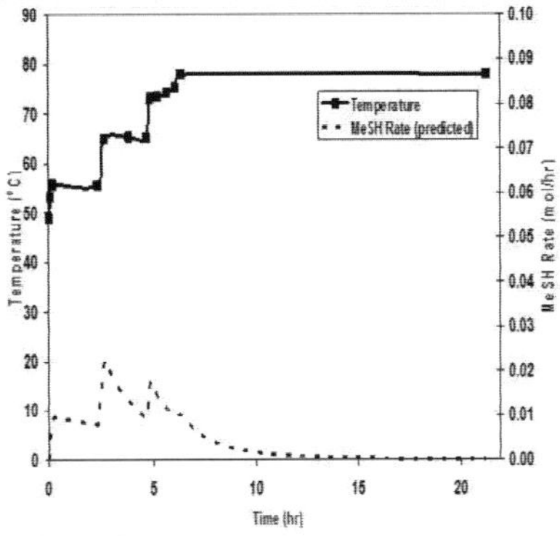

Experimental verification of modeling approach
- **Modeling allows for selection of best temperature profile without extensive experimentation**
- **Tailoring of temperature profiles is possible to accommodate different plant capabilities**

15

Conclusion

Modeling based approach applied successfully to on-scale implementation of MeSH emitting reaction *via*:

- **Optimization of the MeSH generating reaction**
  - Utilized a limited number of experiments to develop and verify a kinetic model for the MeSH generating reaction
  - Developed two reaction temperature profiles which reduced the maximum off-gas flow rate by ~ 60 % while only increasing reaction cycle time by ~ 15 %
  - Reduced redundancy in the scrubber design through an understanding of the off-gas flow rates and ranges expected
- **Interpretation of the on-scale results obtained**
  - Scrubber energy balance provided a view of the reaction progress in the absence of actual samples

25

DR.REDDY'S

...because we care.

Development of Kinetic Model and Process Prediction

Date - 28/04/2011
Venue - DynoChem User Meeting, India
Name - Keerthi Pemula

PSAI: PT- REACTION ENGINEERING GROUP

Development of Kinetic Model and Process Prediction, K Pemula, Dr. Reddy's, DynoChem User Meeting 2011

DynoChem *Dr. Reddy's: Reduce that impurity*

Process/Our system

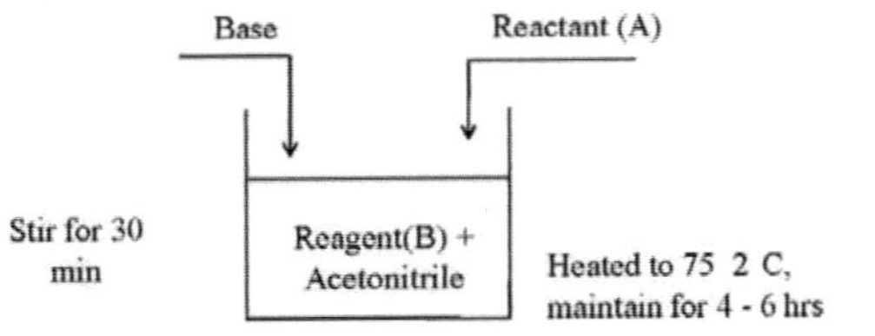

Issue

Excessive formation of Impurity1 (6%), which is difficult to remove and results in yield loss ; Target Impurity1 - 4%

PSAI:PT- REACTION ENGINEERING GROUP

LIFE . RESEARCH . HOPE

Development of Kinetic Model and Process Prediction, K Pemula, Dr. Reddy's, DynoChem User Meeting 2011

DynoChem *Propose mechanism; assess fit*

Mechanism 1

1) A + B	→ Product +C	k_1 = 4.8E-04 L/mol s at 75 C
2) C + B	→ Impurity1 + D	k_2 = 1.7E-04 L/mol s at 75 C
3) C + Product	→ Impurity2 + D	k_3= 1.5E-05 L/mol s at 75 C

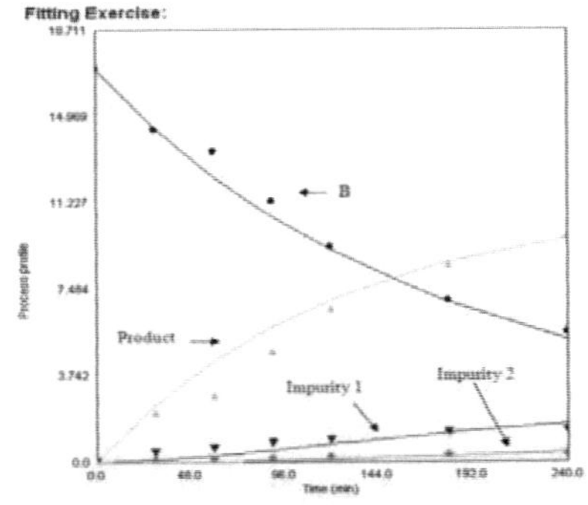

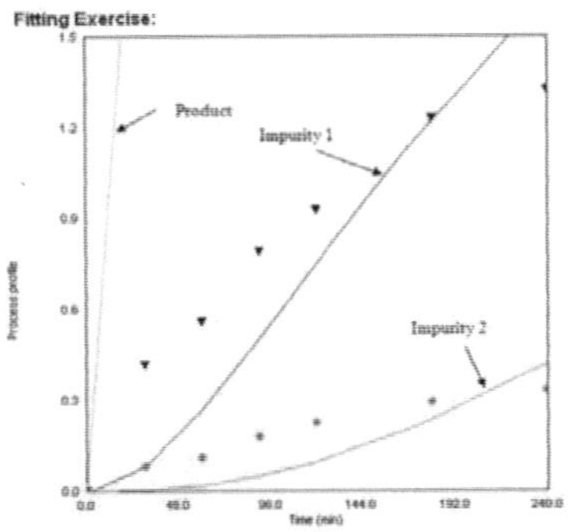

PSAI:PT- REACTION ENGINEERING GROUP

LIFE . RESEARCH . HOPE

DynoChem — *Revise mechanism; assess fit*

Mechanism 2

1) A + B → Product + C — k_1=4.8E-04L/mol s at 75 C
2) C + Base → Intermediate — k_2=1.0E+02L/mol s at 75 C
3) Intermediate + B → Impurity1 + Base + D — k_3=1.7E-04L/mol s at 75 C
4) Product + Intermediate → Impurity2+ Base + D — k_4 =1.5E-05L/mol s at 75 C

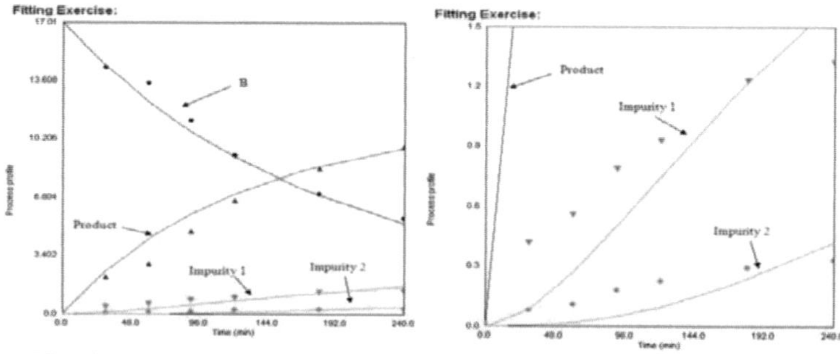

PSAI:PT- REACTION ENGINEERING GROUP

LIFE . RESEARCH . HOPE

DynoChem — *Revise mechanism; assess fit*

Mechanism 3

1) A + B → Intermediate — k_1= 1.5E-04 L/mol s at 75 C
2) Base + B → Impurity1 + E — k_2 = 1.2E-05 L/mol s at 75 C
3) Intermediate → Product + C — k_3 = 1.7E-03 1/s at 75 C
4) Base + Intermediate → Impurity2+ E — k_4 = 7.2E-05 L/mol s at 75 C
5) E + C → Base + D — k_5 = 1.0E+02 L/mol s at 75 C

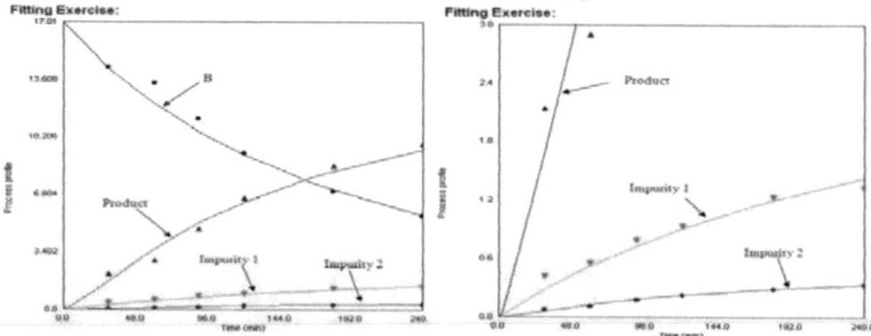

PSAI:PT- REACTION ENGINEERING GROUP

LIFE . RESEARCH . HOPE

DynoChem *Learn the mechanism, unlock success*

Conclusions

- Kinetic curve fitting with Mechanism 3 is perfect for all components
- Final SSQ and rSq are also good for Mechanism 3
- Therefore Mechanism 3 is accepted as a feasible mechanism with the obtained kinetic parameters

Benefits from knowing mechanism

- Changed addition pattern to solvent, Base, A, stir for 30 min and then add B, where A has more selectivity to react with B
- Reduced the quantity of Base from 2 equivalents of B to 0.8 equivalent of B => concentration increased, rate increases and reaction time reduced by 2 hrs
- Impurity1 got reduced by 3%
- Yield improved by 14%

PSAI:PT- REACTION ENGINEERING GROUP | LIFE . RESEARCH . HOPE

DynoChem *Case 2: An impurity is eradicated!*

Process/Our System

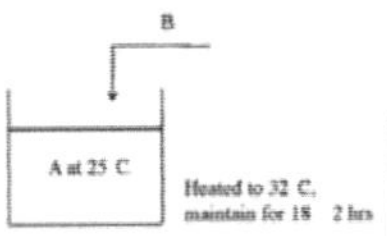

Issue
Reducing the formation of Impurities and increasing yield

PSAI:PT- REACTION ENGINEERING GROUP

Conclusions

- Simulation shows that dilution slows down the reaction
- Optimization shows a dilution with additional 2.5V of solvent at 38.7 C and end time of 20 hrs gives minimum impurities

Actual benefits obtained from knowing the mechanism

- Impurity1, which was difficult to remove is reduced from 0.2% to 0.09%, by reducing reaction time
- Reaction time is reduced by 8 hrs
- Once we came to know that Impurity 2 doesn't form from Impurity1,we tried different ways to isolate it and succeeded in removing it completely
- Since impurities are reduced, by reducing the volumes of solvents in workup, yield was improved

PSAI:PT- REACTION ENGINEERING GROUP | LIFE . RESEARCH . HOPE

Dynochem Modeling for API Process Development:

Early Stage Linking of Laboratory to At-Scale Performance

Jim Marek, Ph.D
Ann Czyzewski, Ph.D
Evelina Kim, Ph.D.
Alex Marchut, Ph.D.
Nandkishor Nere, Ph.D.

Abbott Laboratories,
Process R & D Engineering Department

Abbott
A Promise for Life

DynoChem Modeling for API Process Development: Early Stage Linking of Laboratory to At-Scale Performance, J Marek et. al., Abbott, DynoChem User Meeting 2011

DynoChem *Abbott: CPP solvent swap. Ready, set ...*

What's the Problem?

Jim,

I need some help.
Could you tell me the best way to go from substrate and 40 volumes of Toluene to substrate and 10 volumes of MeOH.
Would like < 1% toluene at the end point.
Want to always have at least 2 volumes of solvent present.
Will be going in the CPP in 2 weeks

Thanks, Rob

Dynochem Modeling for Early Stage API Process Development May 11th, 2011 11 ©2011 Abbott

Abbott
A Promise for Life

DynoChem Modeling for API Process Development: Early Stage Linking of Laboratory to At-Scale Performance, J Marek et. al., Abbott, DynoChem User Meeting 2011

DynoChem *... go? Not so fast ...*

What's the Problem?

12 days later:

30 volumes of 2/1 v/v heptanes-ethyl acetate solvent switch to an endpoint of 99% MeOH (in 5 volumes). I can distill down to ~ 3 volumes.

thanks

Rob

Dynochem Modeling for Early Stage API Process Development May 11th, 2011 12 ©2011 Abbott Abbott A Promise for Life

DynoChem *Answer from solvent swap model*

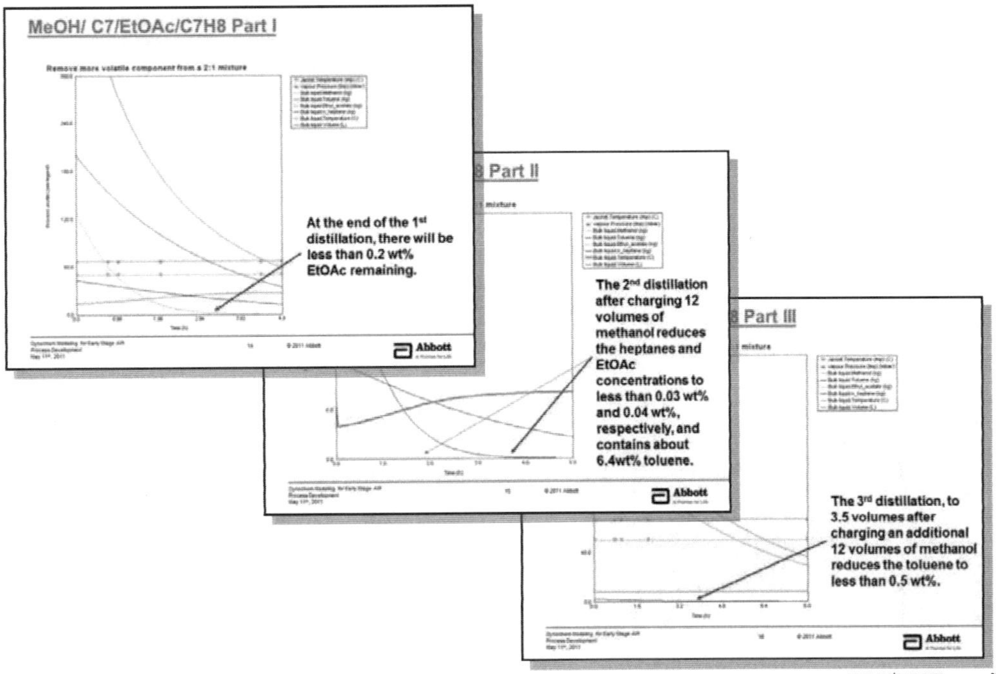

DynoChem *Solvent swap models pay back well*

Distillation

- Solvent Swaps & Chases allow set up for other unit operations, including reactions, crystallizations, extractions.
- Make as efficient as possible to reduce solvent usage, cycle time, and overall costs.
- Optimizing distillations is often overlooked in early and mid-stage API development. This can actually impact the pilot plant efficiency, and number of processes and API's produced.
- One key to maximizing value is getting all functional areas involved.

Dynochem Modeling for Early Stage API Process Development May 11th, 2011 10 ©2011 Abbott Abbott A Promise for Life

DynoChem *Brains + Experiments + Models = Success*

Not so fast sunshine by banjaxx at flickr.com

Hey!

Not so fast ...

Where are the experiments?

DynoChem *Brains + Experiments + Models = Success*

Standing upon the Shoulders of Giants by kjarrett at flickr.com

Someone else did the experiments, long ago!

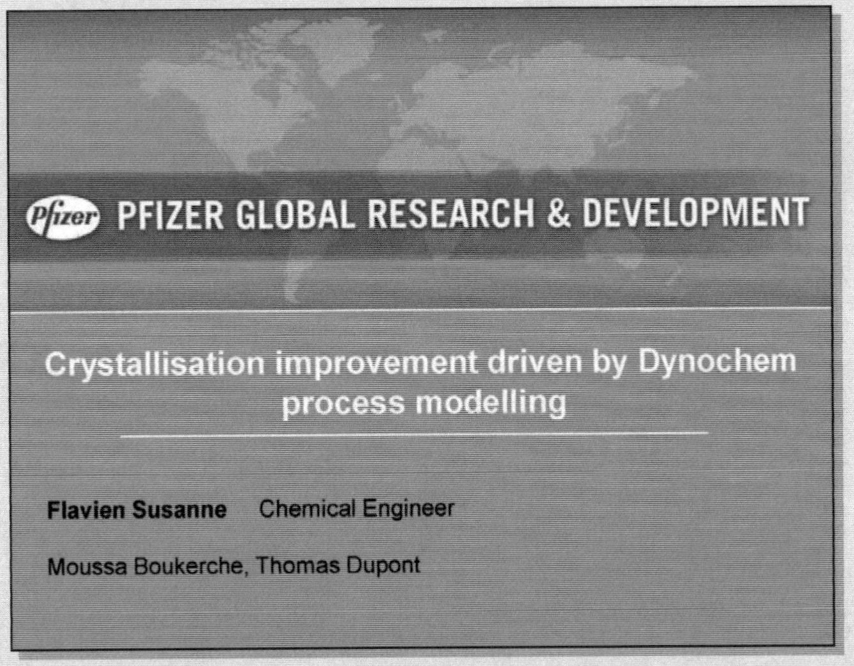

Crystallization Process Improvement Driven by DynoChem Process Modeling, F Susanne, Pfizer, DynoChem Guest Webinar 2011

DynoChem — *Pfizer: Crystal size 'roller coaster'*

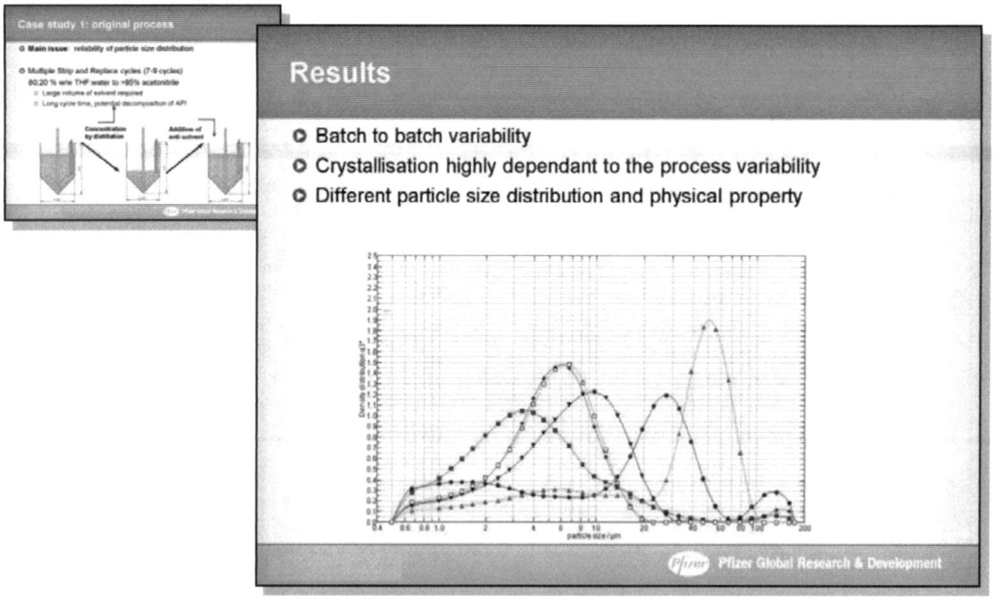

[*Crystallization Process Improvement Driven by DynoChem Process Modeling, F Susanne, Pfizer, DynoChem Guest Webinar 2011*]

DynoChem — *Modeling alongside experiments*

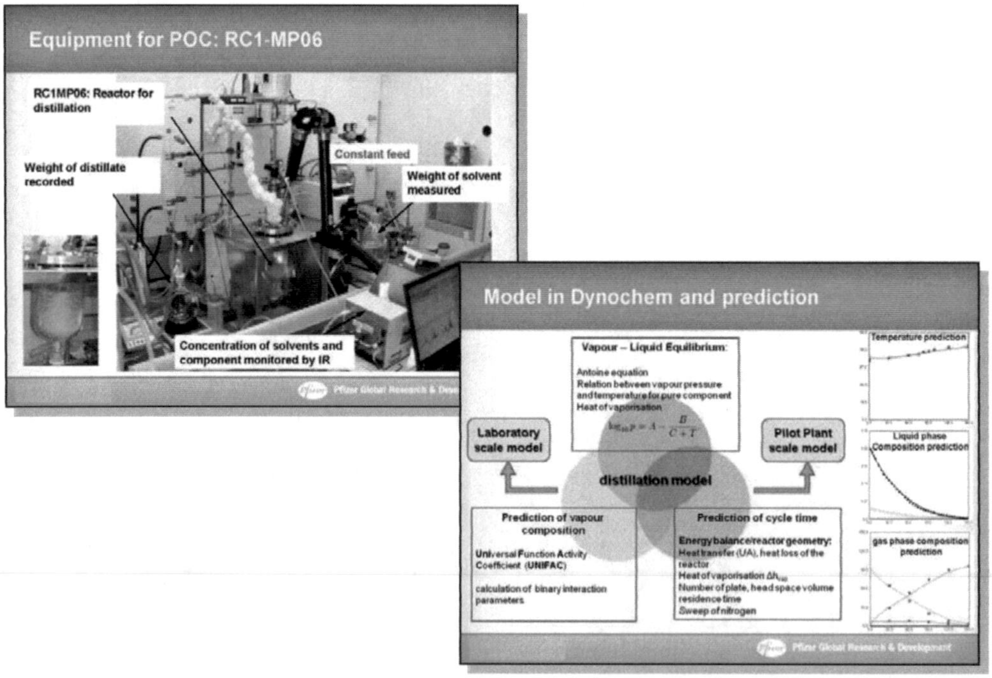

DynoChem *Marry model and equipment data*

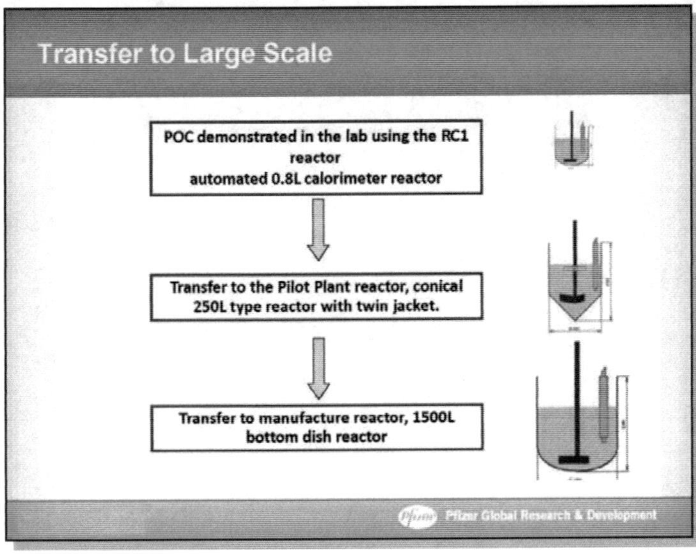

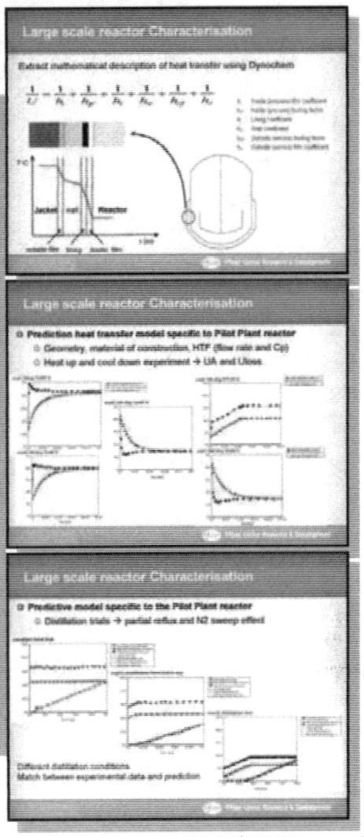

DynoChem *Pfizer: Modeling improves repeatability*

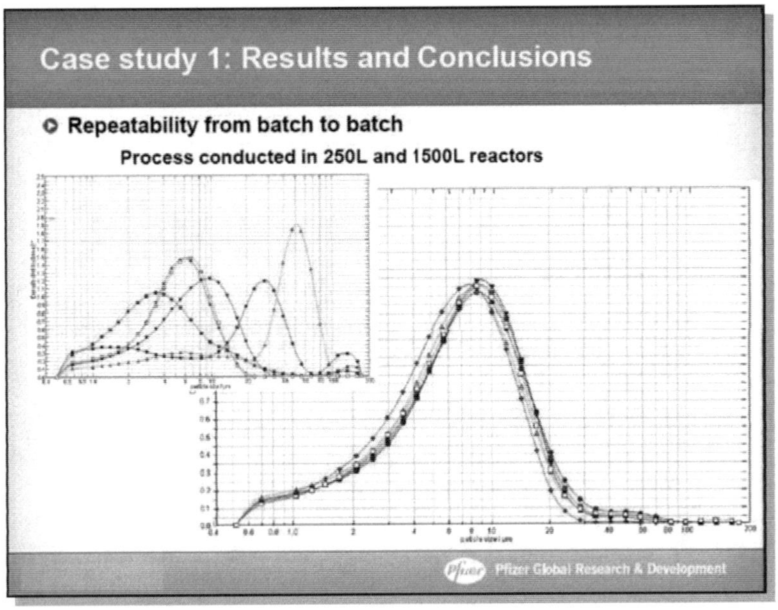

AMGEN

Pioneering science delivers vital medicines™

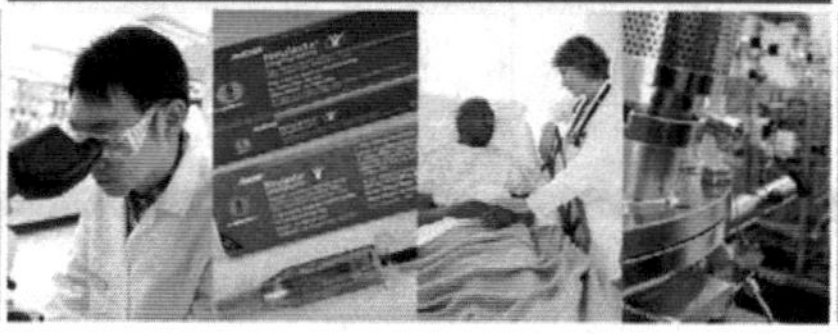

Applications of the Simple Filtration and Centrifugation Tool

William Trieu

10-May-2011

Applications of the Simple Filtration and Centrifugation Tool, W Trieu, Amgen, DynoChem Guest Webinar 2013

DynoChem *Amgen: Solid-liquid separation*

Pressure Filter Setup

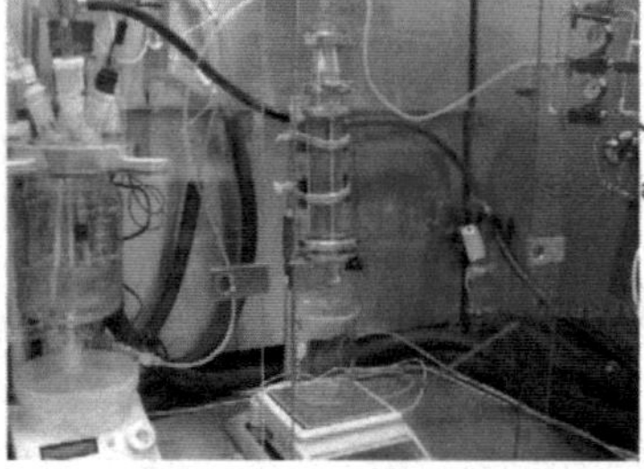

Dynochem Output

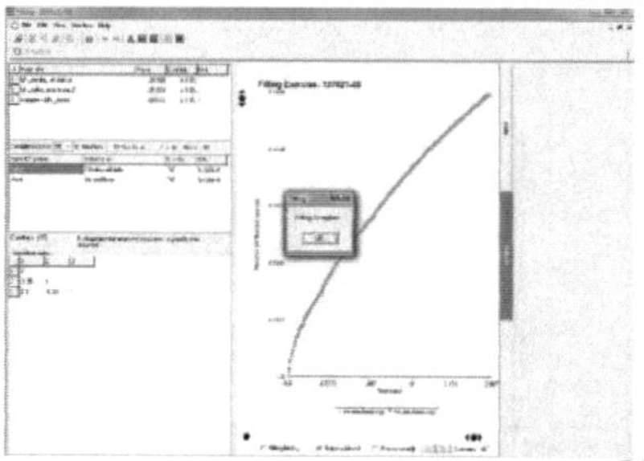

Modeling alongside experiments

DynoChem *Push, spin or back to the lab?*

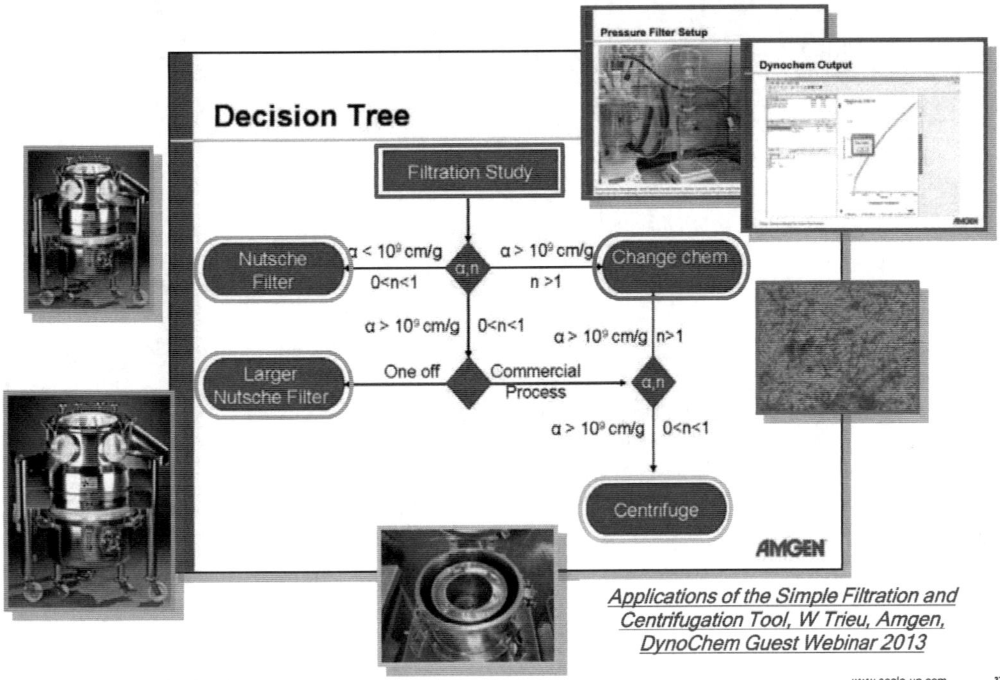

Applications of the Simple Filtration and Centrifugation Tool, W Trieu, Amgen, DynoChem Guest Webinar 2013

Bristol-Myers Squibb

Application of DynoChem® Reaction Modeling to Quality by Design

Daniel Hallow, Boguslaw Mudryk, Alan Braem,
Justin Burt, Lucius Rossano, Srinivas Tummala

Bristol-Myers Squibb
Chemical Development

DynoChem® Webinar
July 24th, 2013

Application of DynoChem Reaction Modeling to Quality by Design, D Hallow, BMS, DynoChem Guest Webinar 2013

DynoChem — *BMS: Define design space; QbD filing*

Critical Impurity Formed in the Penultimate Step

Critical Impurity:

Impurity "C" — *Penultimate Step*

Impurity "D" — *API Step*

Goal

- Define reaction mechanism and reaction rate kinetics
- Develop a mechanistic model to predict the impurity formation
- Define a design space for the penultimate step to ensure control of impurity within specifications
- Incorporate the developed mechanistic model and into the design space in regulatory filings for approval

Bristol-Myers Squibb

Application of DynoChem Reaction Modeling to Quality by Design, D Hallow, BMS, DynoChem Guest Webinar 2013

DynoChem — *Develop mechanistic model*

Chemistry/Modeling Equations

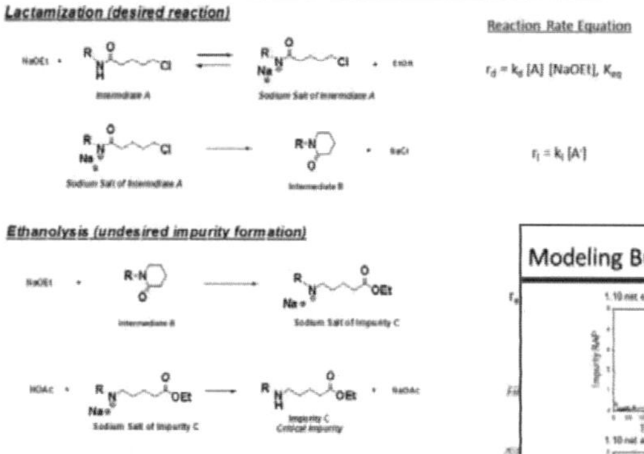

Modeling Building Experiments

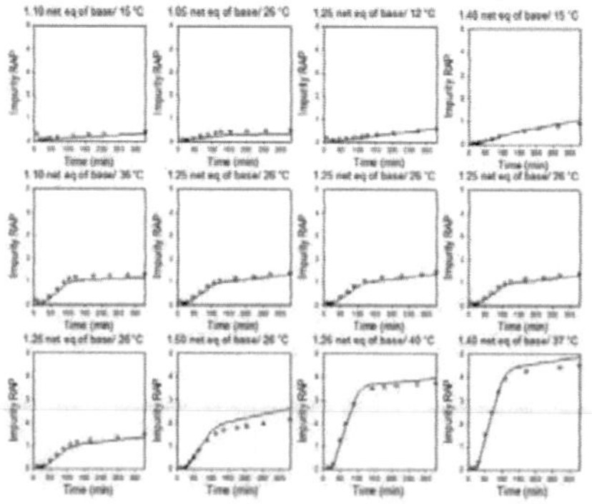

Bristol-Myers Squibb

DynoChem — *Explore key responses*

Design Space Exploration

- DynoChem® design space exploration tool used to simulation numerous experiments (~80) to generate response surfaces
- Easily visualize the effect of key parameters on substrate and impurity levels across the parameter space:
  - High base and temperature – reduce residual substrate
  - Low base and temperature – reduce impurity formation

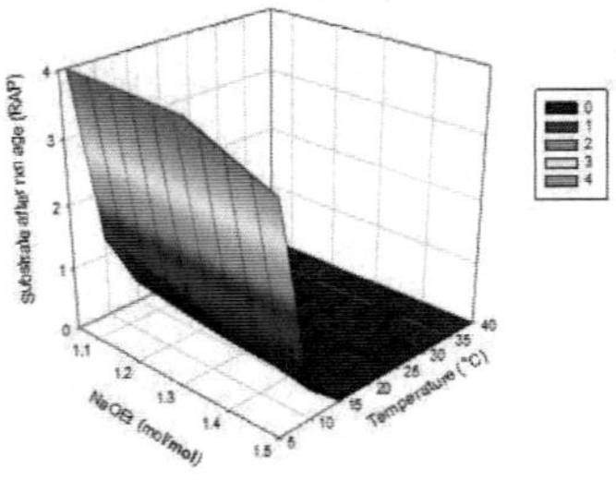

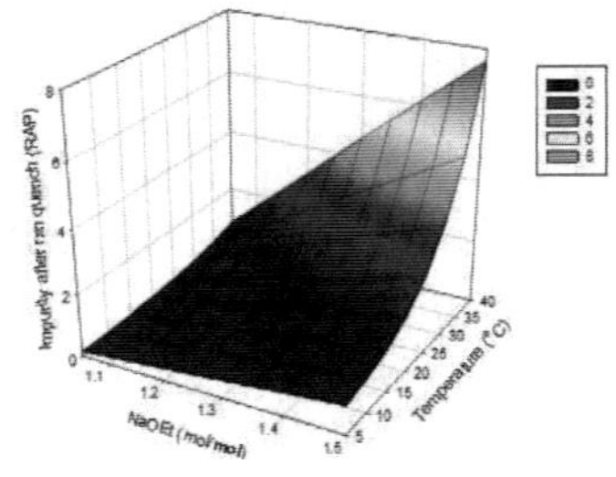

DynoChem — *Identify design space, verify*

Design Space Verification

- Diagonal edge represent highest-risk region of the design space
- Experimental Testing - Demonstration of process parameters that provide assurance of quality
  - Corners of the design space
  - Experiments along diagonal edge
  - Four experiments at one point to demonstrate reproducibility

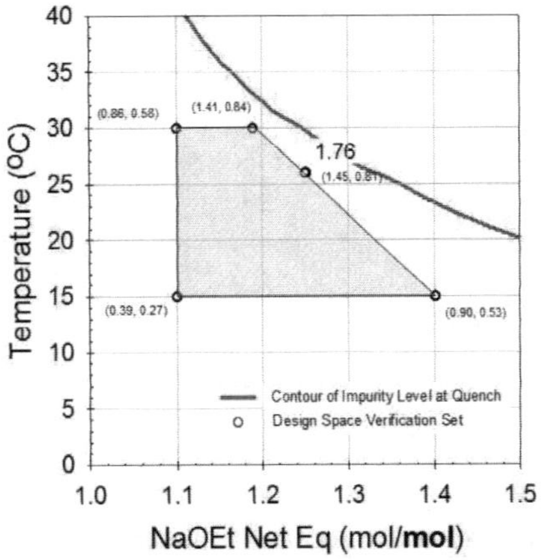

Bristol-Myers Squibb

DynoChem *BMS: Design space, QbD filing, approved*

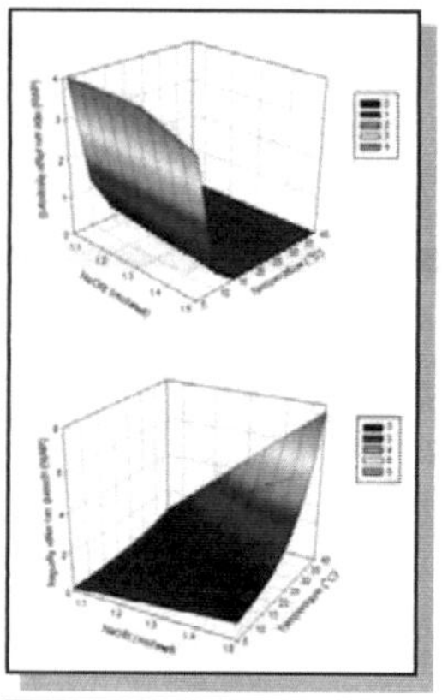

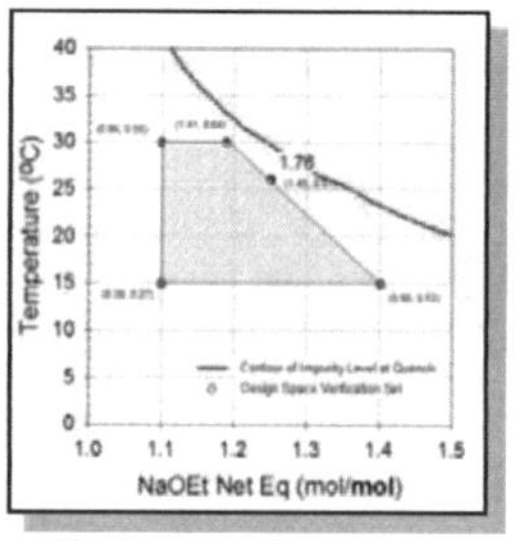

Conclusions

DynoChem® Utilization

- Fitting Tool used to fit kinetic parameters for four reactions
  - Model to predict impurity profile
- Simulator used to understand the effect of process parameters
  - Understanding of effect of base, temperature, time, solvent composition
  - Suggested new temperature profile to increase process robustness
  - Assessment of risk through out the design space
- Design Space exploration used to conduct numerous simulations that were summarized in response surfaces

Quality by Design

- Mechanistic model served as a guide to define a design space
- Design space was a simple polygon shape in a reduced parameter space
- Design space verification was completed through experimental testing

Design space submitted as part of regulatory filing under QbD. Approved.

Bristol-Myers Squibb

Application of DynoChem Reaction Modeling to Quality by Design, D Hallow, BMS, DynoChem Guest Webinar 2013

www.scale-up.com 43

DynoChem *GSK: Kinetics + DoE is best*

Previous : 88% yield, not robust (initial DoE)

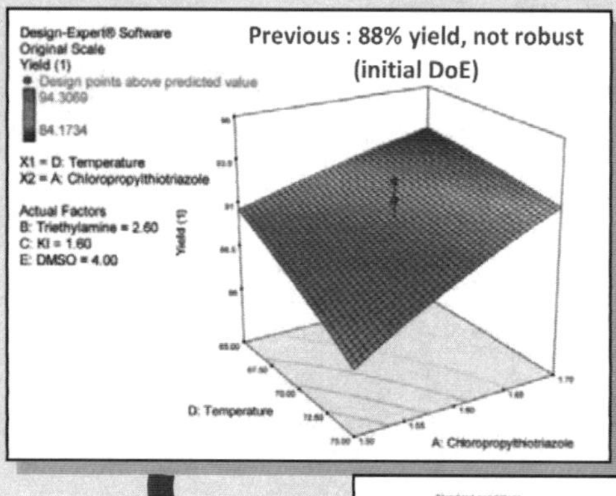

New: 94% yield, more robust (in-silico DoE + Kinetics)

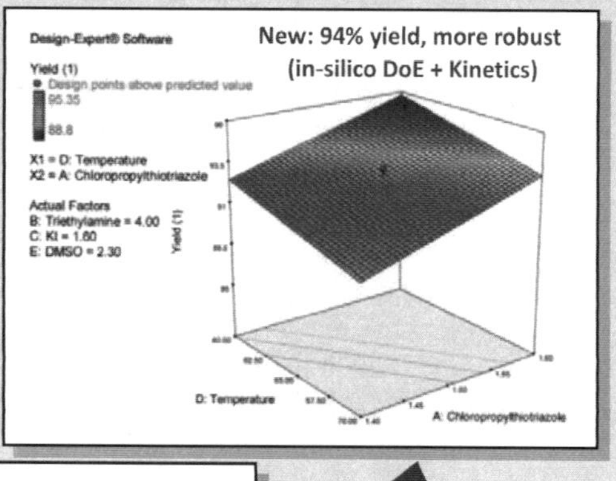

Kinetic model predictions and experimental data

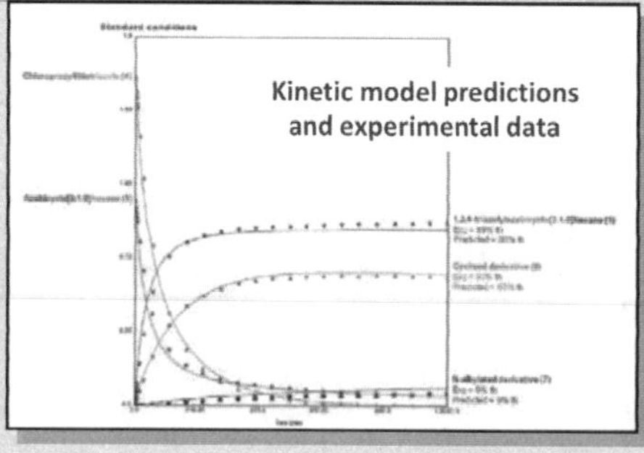

A Mechanistic Insight into a Simple C-N Bond Formation via SN2 Displacement: A Synergistic Kinetics and Design of Experiment Approach, L Massari, et. al., OPRD, 2010, 14, 1364-1372

GlaxoSmithKline

www.scale-up.com 44

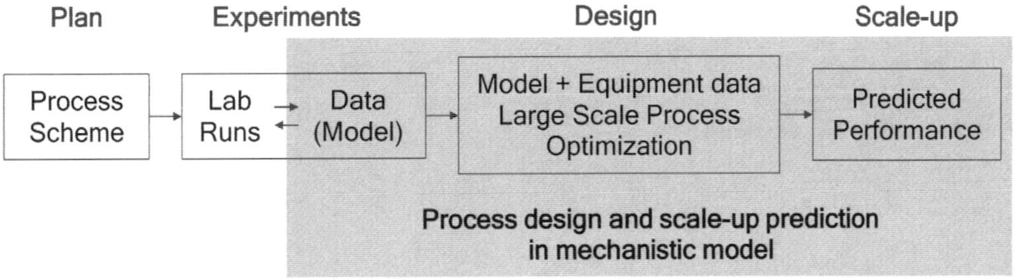

Cows along the path by *Jon Fife* at *flickr.com*

DynoChem *The fast track to process success*

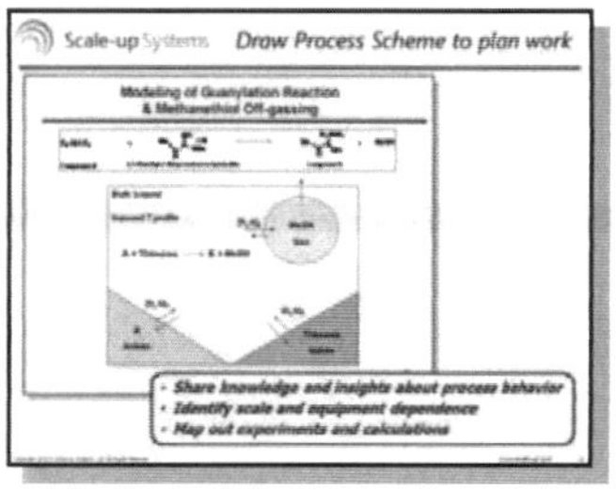

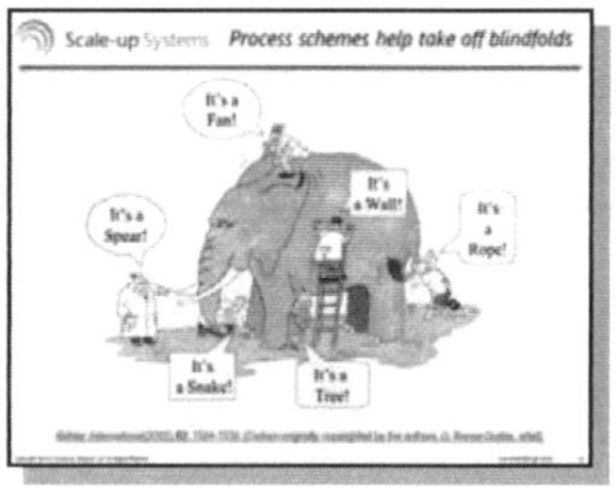

Draw a process scheme to promote collaboration, see scaling concerns and plan work

DynoChem *Webinar on using Process Schemes*

Andrew Cosbie
Process Engineer at Amgen
Ventura, California | Biotechnology
Current Amgen
Education University of California, Santa Barbara
Send a message

Process Schemes: Building Value into Process Understanding and Engineering Assessments

Wednesday, June 18

Register now at http://scale-up.eventbrite.com/

DynoChem

The fast track to process success

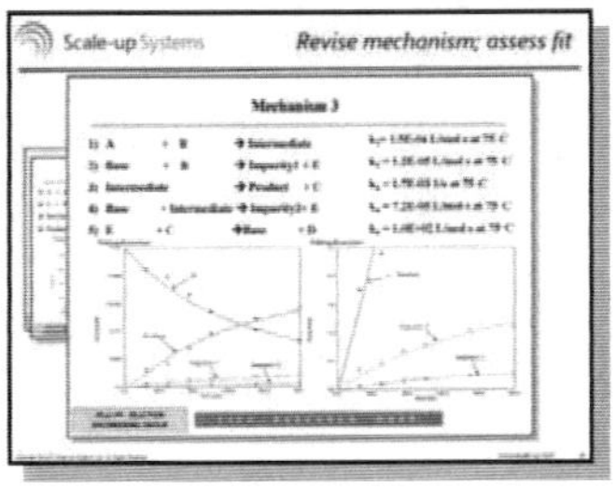

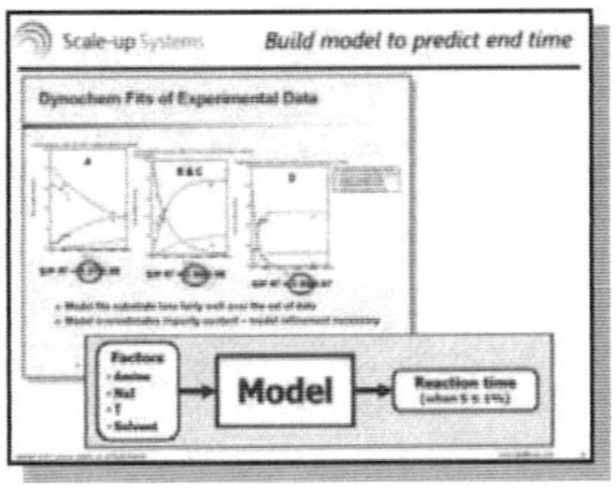

Draw a process scheme to promote collaboration, see scaling concerns and plan work

Perform experiments to understand the process, not to optimize it

DynoChem

The fast track to process success

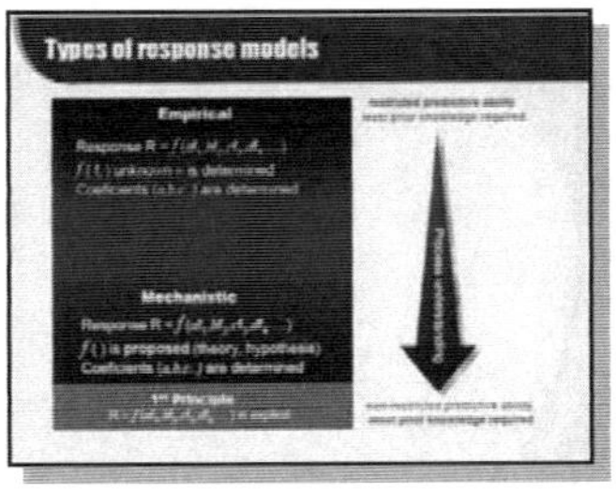

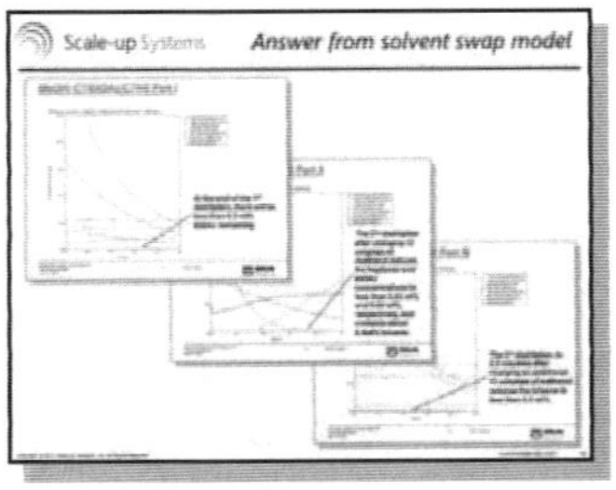

Draw a process scheme to promote collaboration, see scaling concerns and plan work

Perform experiments to understand the process, not to optimize it

Capture understanding in a mechanistic model for trustworthy predictions

DynoChem *The fast track to process success*

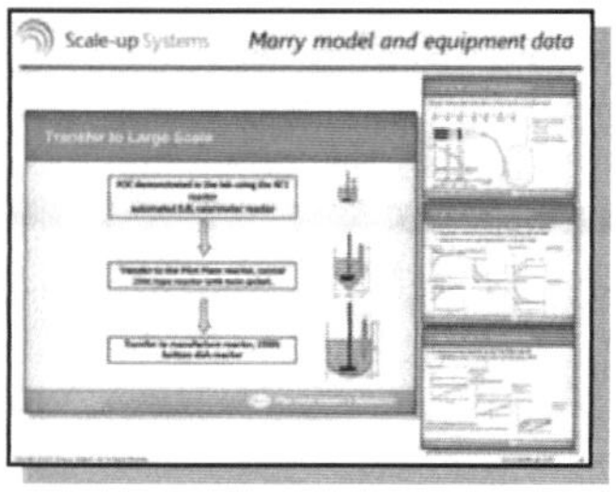

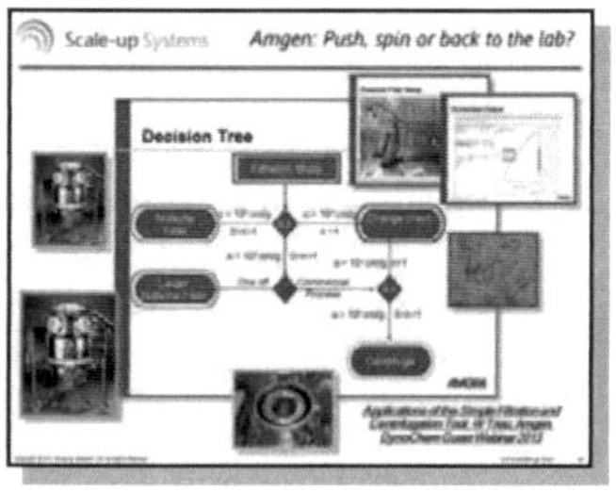

Draw a process scheme to promote collaboration, see scaling concerns and plan work

Perform experiments to understand the process, not to optimize it

Capture understanding in a mechanistic model for trustworthy predictions

Marry models and equipment data to increase scale-up confidence

DynoChem *The fast track to process success*

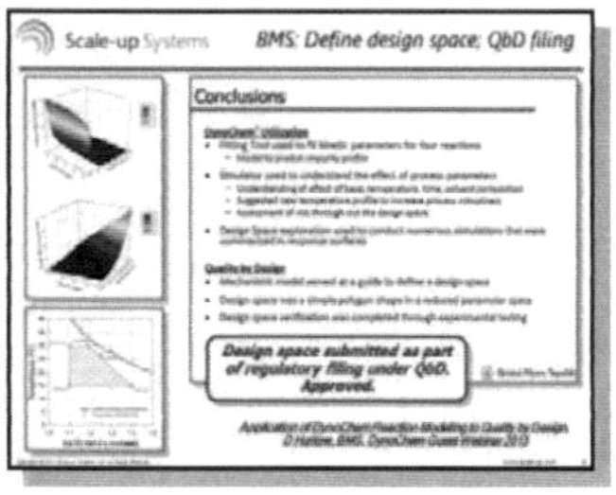

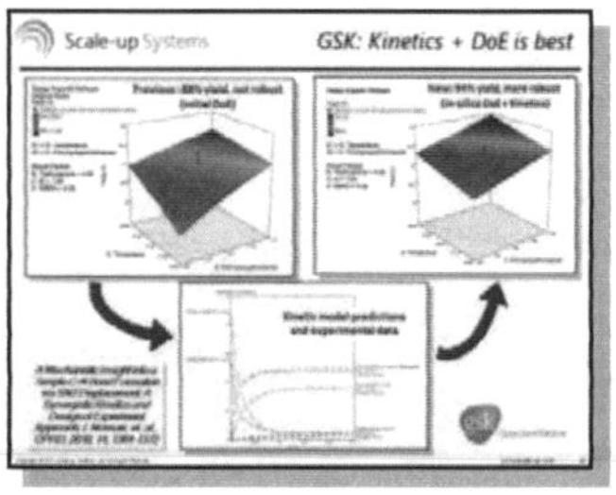

Draw a process scheme to promote collaboration, see scaling concerns and plan work

Perform experiments to understand the process, not to optimize it

Capture understanding in a mechanistic model for trustworthy predictions

Marry models and equipment data to increase scale-up confidence

Construct design spaces and submit regulatory filings on a solid mechanistic foundation

DynoChem

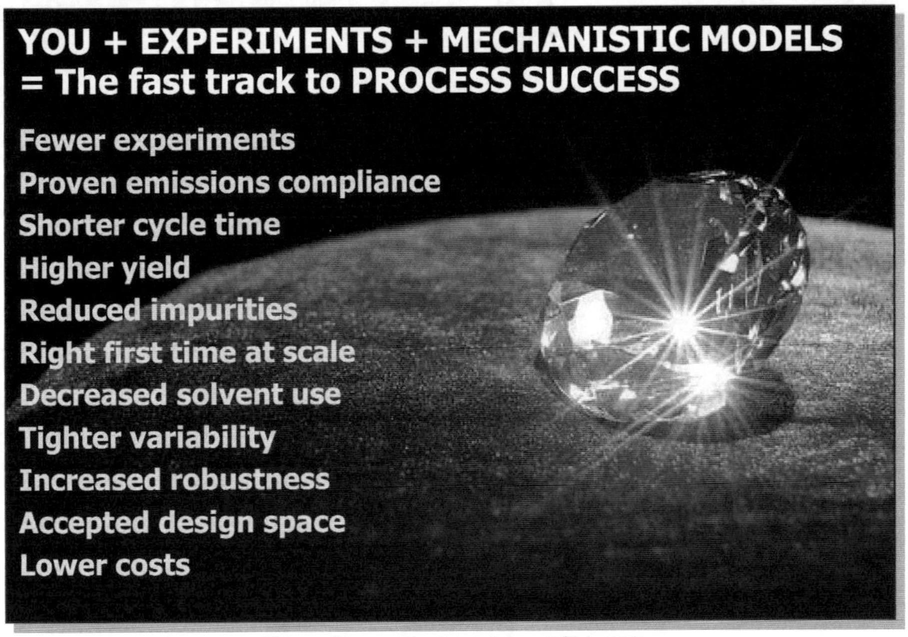

Diamond in my eyes by racineur at flickr.com

Scale-up Systems — *Acknowledgements*

The many people whose stories I shared with you

My colleagues at Scale-up Systems (alphabetically)

Andrew Bird

Steve Cropper

Joe Hannon

Wilfried Hoffmann

Thank you for your time today

To learn more ...

See more case studies in the Appendix

Visit www.scale-up.com and

Visit dcresources.scale-up.com for more cases, ...

Talk with us ... Brains + Experiments + Models *for you*

NALAS
Engineering Services, Inc.

Filtration and Drying Operations: Experimental and Modeling Techniques to Preserve Particle Properties upon Scale-up to Agitated Filter Driers (AFD's)

Matthew Jörgensen, Jerry Salan, David am Ende
AIChE PDS 6/11/2014

1 SOLUTIONS THAT SCALE

Outline:

- Who is Nalas and what we do
- Advantages/Issues with isolation using Agitated Filter Driers (AFD)
- Process Evaluation for Isolating an API
  - Characterizing Filtration
  - Evaluation of the risk to form agglomeration
- Modeling of AFD and Mitigation strategies to avoid agglomeration

2 NALAS Engineering Services, Inc.

Solutions that Scale!

Mission

Our mission is to promote technological innovation in the chemicals industry by applying science-based knowledge to solve the problems and achieve the goals of our CUSTOMERS. The creative ideas conceived by the customer will be realized by the Nalas team of scientists and engineers. We will support our customers with all aspects of process engineering that result in a new dimension of performance and enable SOLUTIONS THAT SCALE from the laboratory to commercial production. We will provide sustainable and green processes which can be transitioned to industry, thus enabling lowest-cost domestic production of chemicals critical to our NATION'S military, industrial, and pharmaceutical needs.

3 NALAS Engineering Services, Inc.

Engineering Services

utions | Services | Capabilities

Chemical synthesis
We have experience working with a large number of reaction steps and subsequent workup and isolation stages to identify potential problems in scale-up to pilot plant and production campaigns. >> More

Testing
We have a wide array of testing equipment, ranging from purity measurements, thermal analyisis, particle characterization, and small-scale sensitivity for energetic materials. >> More

Safety
Process Safety is critical to any chemical process and the proposed facility, whether you are synthesizing just a few kilograms or planning production of metric tons. >> More

Products
A full list of our available chemical products.>> More

Modeling
We are proud users of process models such as DynoChem® and Visimix® to precisely determi potential scale-up failures. >> More

Training / Webinars
We offer training and seminars to help familiari with tools such as automated laboratory reactor real-time heat flow analysis, and other process analytical tools >> More

Solid State Chemistry & Characterization
As experts in solid state chemistry, we offer cry cocrystal and polymorph screening and characterization. >> More

Reaction Monitoring
Reaction monitoring is a defining capability that Nalas apart from everyone else in the industry.: More

Broad range of services & capabilities supporting chemical process development…

http://nalasengineering.com/

4 NALAS Engineering Services, Inc.

Nalas Functionalities at a Glance

- Process Development
- Chemical Synthesis
- Chemical Engineering
- Reaction kinetics and Modeling
- Analytical Services
- Solid-State Chemistry
- Single Crystal and Powder Xray
- Crystallization Design
- Process Safety
- Process Design
- Scale-Up
- Kilo-lab Manufacturing
- Testing Services

5 NALAS Engineering Services, Inc.

Agitated Filter Dryers (AFDs) – Some Background

Why Do We Love Them?

Filtration, Washing, and Drying All in One!
Excellent Containment
The Swiss Army Knife of isolation/drying.

Why Don't We Love Them?

Independent control of attrition, agglomeration, and drying behavior is very difficult if not impossible.

6 NALAS Engineering Services, Inc.

The Problem(s): Prediction of Material Behavior and Performance upon Scale-up

Desire is that AFD will not significantly change the desired particle size distribution, quality, morphology, or form.

Input Slurry of Intermediate or API →

→

But there is currently a lack of Understanding! We cannot currently predict apriori the output particle size distribution from our AFD's including validation batches...What is the cost of failing validation or having to rework...??

- Attrition of Particles
- Agglomeration
- Balling
- Long cycle times
- Can lead to:
  - *Out-of-spec material*
  - *Impact on Drug Product*
  - *Validation delays*
  - *Higher Costs*

7 NALAS Engineering Services, Inc.

Agglomeration in AFD's

- Balling is most often seen in agitated driers and is generally worse in agitated filter driers (BMS).
- Agglomeration is often caused by agitating the cake too aggressively while it still contains a significant amount of solvent.
- Avoid "balling" by avoiding agitation of wetcakes above the critical moisture content - ie during initial stages of drying – ensure the nitrogen blow through achieves LOD below critical moisture content
  - (We may need to use DOE's in combination with other techniques to examine sensitivity of LOD on agglomeration)
- Solubility of the product in the remaining solvent mixture on the cake is also a potential contributor– as the cake dries a solvent could be enriched creating higher concentrations of product and eventually crystal bridges fusing into agglomerates.
- Understand the solubility and solvent drying profiles to ensure this method of agglomeration is minimized. This mechanism is similar to wet granulation where a solvent is sprayed on to API and excipients, typically under high agitation, and dried to achieve desired agglomerates of desired properties.
- Design:
  - Wash solvent selection is important,
  - Select optimum temperature and pressure of deliquoring/filtration
  - understanding the drying profile for mixed solvents
  - Affect of moisture content on agglomerate formation in lab AFD

8 NALAS Engineering Services, Inc.

Types of Agglomeration

- Agglomerates formed during agitated drying
  - "balling" or agglomerates that form via capillary forces, during consolidation of damp powder, held together by crystalline bridges once the solvent is evaporated away.
- Agglomerates that form during compression of filter cake such as heel or resulting from non-uniform wash.
- Cake hardening
  - e.g. during static or tray drying
- Agglomerates formed during crystallization

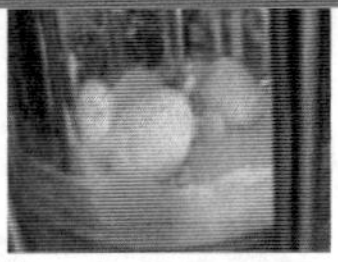

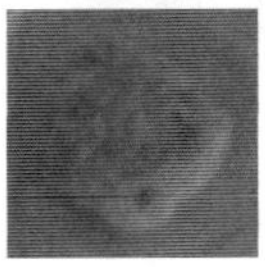

am Ende, D.J., Birch, M., Brenek, and M. T. Maloney, Development and Application of Laboratory Tools To Predict Particle Properties upon Scale-Up in Agitated Filter-Dryers, *Organic Process Research & Development* 2013 *17* (10), 1345-1358

9 NALAS Engineering Services, Inc.

Sensitivity to Agglomeration/Attrition/balling?

- No: Then product output from AFD should not be sensitive to stirring profile as a function of moisture content
  - Ie it should not matter how/when it is stirred during drying
- Yes: Then we need to specify detailed drying and stirring strategy specific to the potential failure modes.

10 NALAS Engineering Services, Inc.

Case Study illustrating Nalas Approach:

- Customer has a Sodium Salt API monohydrate
- Isolation Unit-Op consists of filtering and drying material in AFD
- Need to dry without losing the monohydrate form
  - Dehydrates under N_2
- Would like to dry as quickly as possible under vacuum in AFD
- Need to identify protocol and agitation approach to prevent agglomeration in the drying step

11 NALAS Engineering Services, Inc.

First steps, Evaluation of Filtration

- Can the Filtration be optimized to better de-liquor the cake?
- Is the cake compressible?

The more solvent that is removed during the Filtration step, the less solvent that needs to be dried.

12 NALAS Engineering Services, Inc.

Filtration Data: Ethanol at 20°C

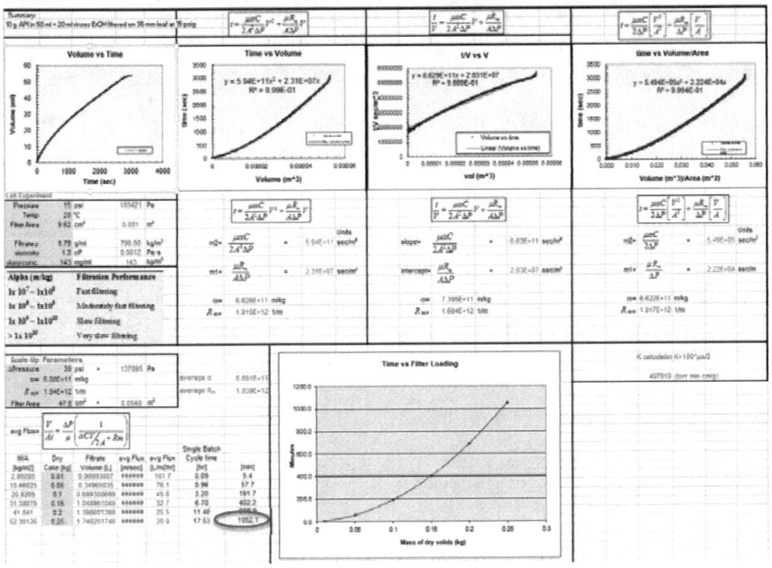

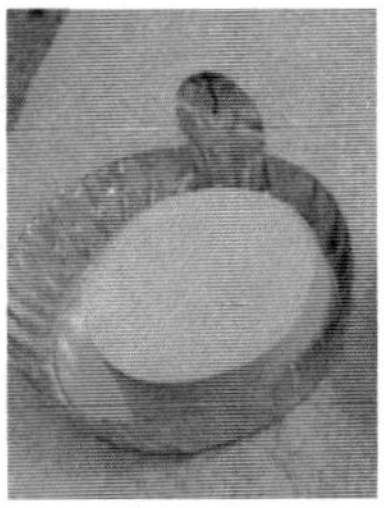

Cake Hardens On Drying

13 NALAS Engineering Services, Inc.

Cake Compressibility:

Evaluation of filtration at 3 pressures allows for determining cake resistance as a function of pressure

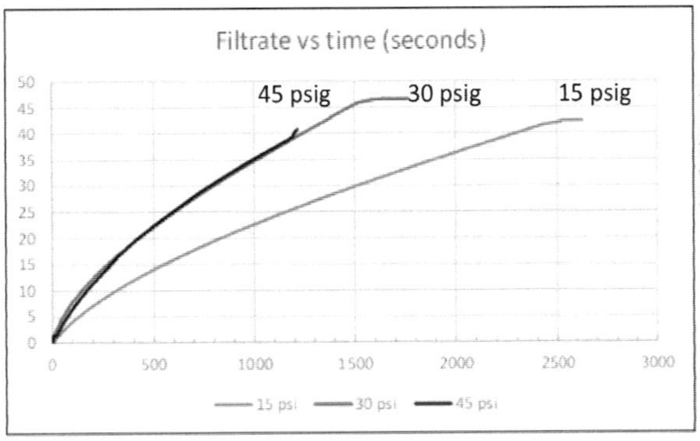

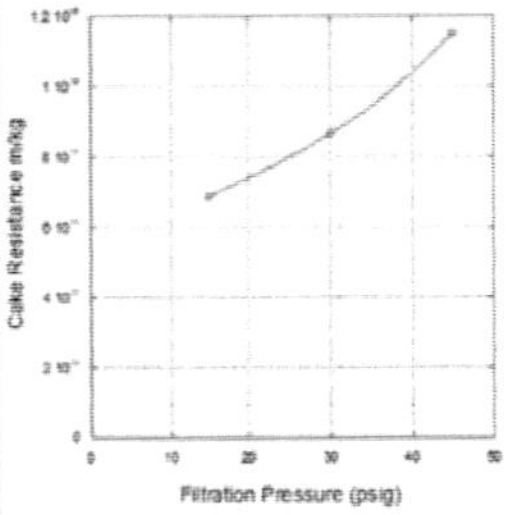

Optimal Pressure for Rate and % LoD

To minimize N_2 Filter under vacuum But use N_2 pulse at 30 psig

Pressure psig	Cake Resistance (x10[11]) m/kg	Wet-Cake % LoD (at the end of Filtration)
15	6.9	38% (0.61 g/g API)
30	8.6	31% (0.45 g/g API)
45	11.5	35% (0.54 g/g API)

At end of filtration the LOD is still over 30%

14 NALAS Engineering Services, Inc.

Rheomtery/Sticky point test

- Drying can be thought of as the reverse of a wet granulation process
- In wet granulation, solvent (normally water) is added to a mix of dry powders to cause agglomeration
- In drying, solvent is removed from a wet powder, normally trying to avoid agglomeration
- Agglomeration tendency ('wet mass consistency') is controlled by 3 main factors
  - Solvent content
  - Mixing speed (energy input)
  - Mixing time

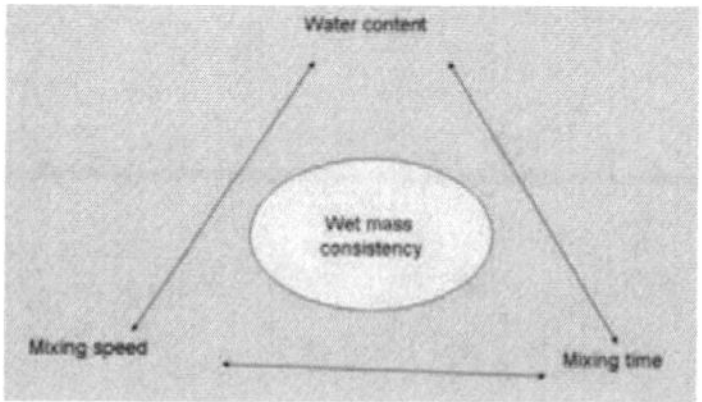

'Granule' development

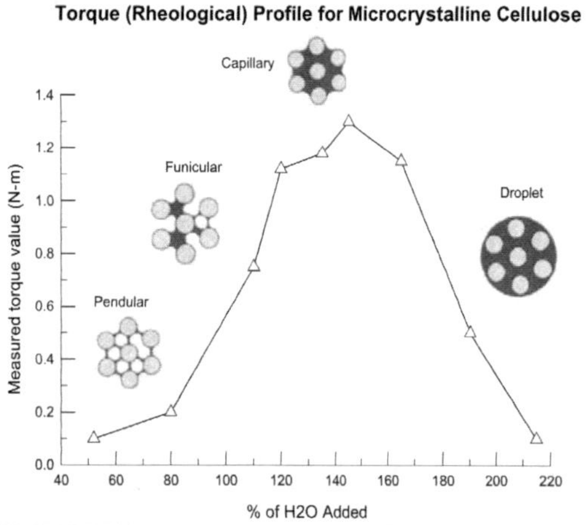

Parker, Rowe and Upjohn, Pharm. Tech. Int. (1990) 9: 50-64

- If energy is put in at the capillary stage particles are forced closer together which increases the capillary force, holding the agglomerate together more tightly
- If drying is applied when agglomerates like this have formed there is the risk of 'drying in' by formation of crystalline bridges due to solvent evaporation

Caleva Mixer Torque Rheometer

Torque

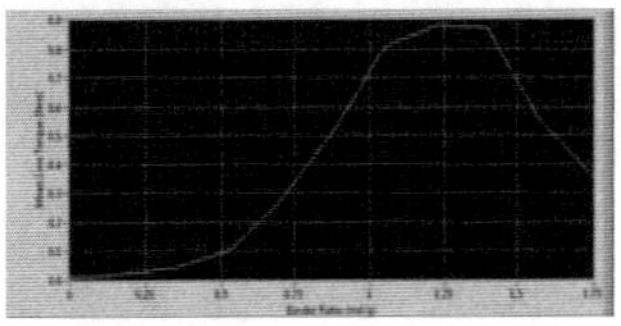

Binder ratio or moisture content

ml solvent/g powder

Load Cell corresponds to torque reading

Mixing Chamber

Measures the stickiness of the wet-cake

Determination of 'sticking point' (agglomeration) using Caleva Mixer Torque Rheometer

Agglomeration tendency increases with increasing solvent content.

- Leading to a measurement of increased torque for agglomerating ('sticky') mixtures

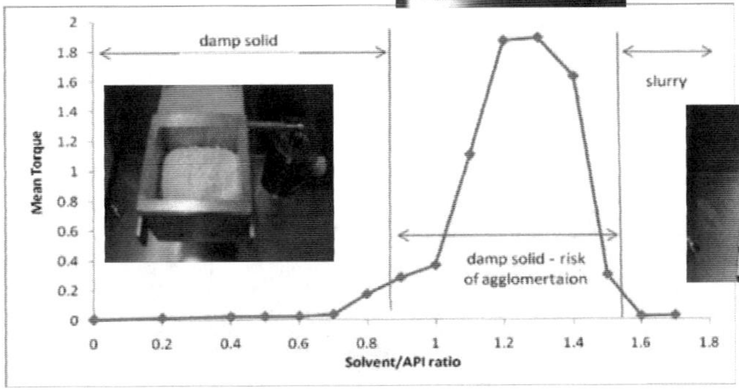

Benefit of Partial Drying/Extended Blow-through before agitation: Avicel PH102A Water Wet

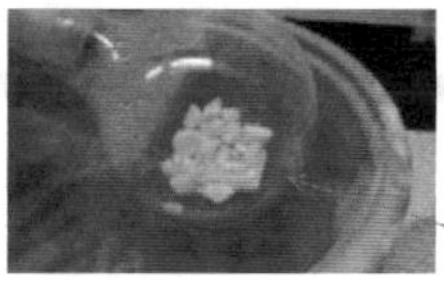

Following Filtration of Avicel /Water Slurry, The Moisture content was measured to be 50% wet (i.e 1 ml/g dry solid). Tumble dried in a rotovap. This cake results in dried large agglomerates

If the cake is first blown with Nitrogen to reduce the Moisture content down to 15% And then tumble dried, Initial wet lumps break up to Free flowing powder

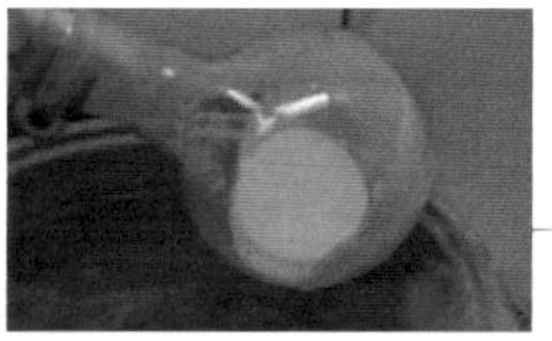

Mixer Torque Rheometer

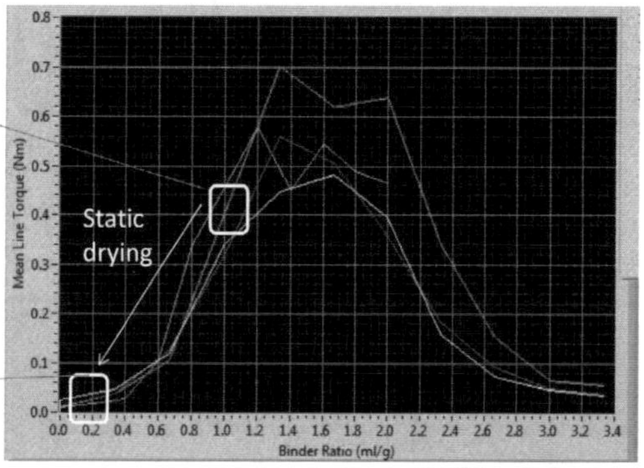

Assess Risk of Agglomeration During Agitated Filter Drier Operation : Sodium Salt API monohydrate

- Develop a drying protocol to minimize agglomerates
  - 5-10 micron particles
  - Wash Solvent EtOH
  - Dehydrates under N_2
- Approach:
  - Need lower moisture wet-cake before agitation (avoid squeezing cake)
  - Improve deliquoring – optimize filtration conditions
  - ~~Use Nitrogen Pressure pulses~~
  - ~~Centrifuge if available~~
  - Examine alternative wash solvents
  - Static Dry until critical moisture content reached near 24 %.
  - Modeling / Predict Drying Performance to identify agitation strategy

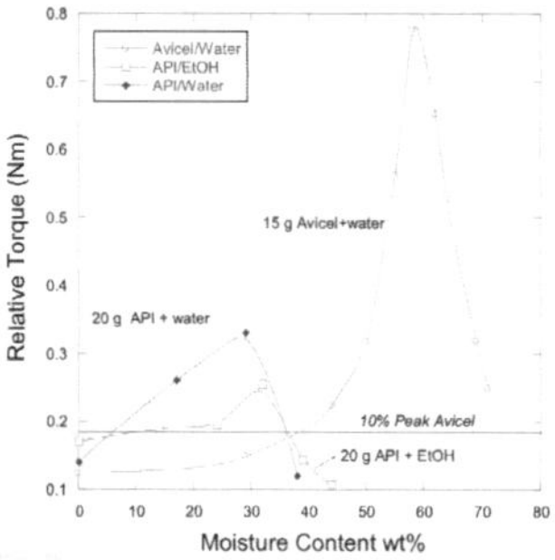

Dry Powder 24% wet 32% wet 39% wet 49% wet

Resonant Acoustic Mixing: Solvent Screen

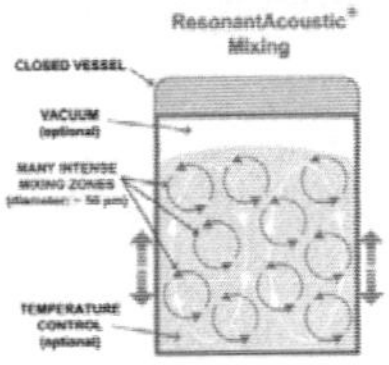

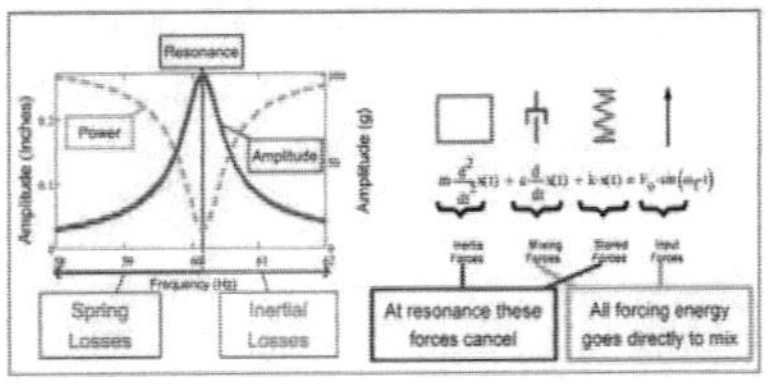

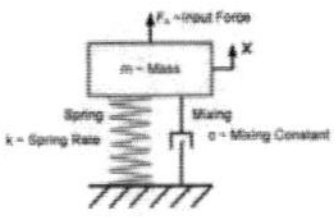

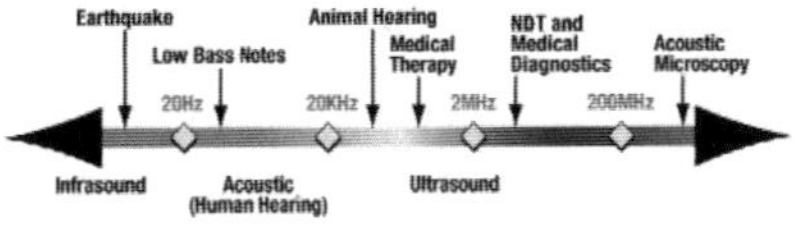

Resonant Acoustic Mixers Operate near 60 Hz (Acoustic Region) and can impart between 0 to 100 G's

21

Resonant Acoustic Mixing for Agglomeration Solvent Screen

- The LabRAM Screen is a much simpler version adapted from an approach previously reported in the litrature[1]
- Using the LabRAM it is possible to Rapidly test multiple solvent at the small scale
  - Insures representative wetting of the solids even with 0.5ml of solvent.
  - With multi vial holder, six solvents can be screened at once
  - Results are a yes/no and helps to down select solvents for MTR testing to determine critical moister content

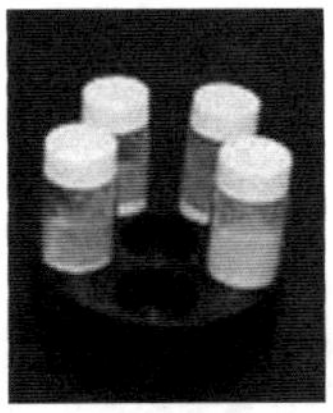

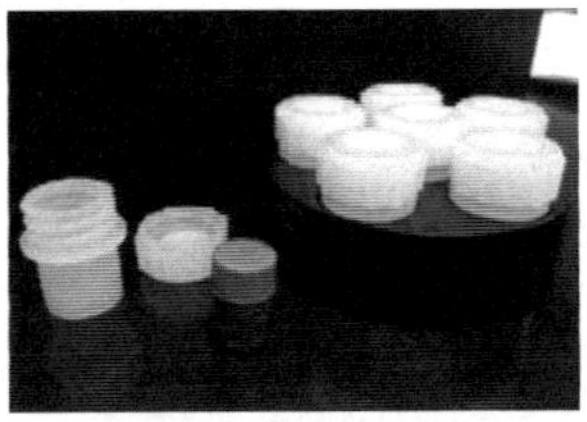

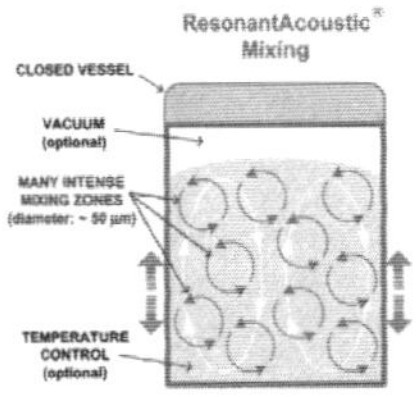

[1] Zang, S, Lamberto, D. J. "Development of New Laboratory Tools for Assessment of Granulation Behavior During Bulk Active Pharmaceutical Ingredient Drying" J. Pharm Sci. 2014 Jan, 103 91): 152-60

22

Agglomeration Solvent Screen

Procedure

- 1 g API per vial
- Add 0.4 g (0.5 ml) of solvent ≈ 30 wt% wet-cakes
- Mix in RAM for 3 min at 30% intensity
- Loosen hard-packed cake
- Mix in RAM for 3 min at 30% intensity
- Agglomerate yes or no?
- Dry

23 NALAS Engineering Services, Inc.

Solvent Screen Observations

Solvent	After First 3 min of mixing	After Second 3 min
Ethanol	sticky	Large "boulder"
Toluene	Hard packed	Hard Cake medium size agglomerates, hard lumps
EtOAc	Balls, chunks	Hard cake packed difficult to dig, large agglomerates
Acetone	No significant agglomerates	A couple of spherical agglomerates softer than from EtOAc
MTBE	No significant agglomerates	Soft lumps
THF	A few spherical agglomerates, hard cake	A few small spherical agglomerates but easier to break
Heptane	Loose cake	Minimal agglomeration, loosely packed, lumps soft

24 NALAS Engineering Services, Inc.

Solvent Screen for Agglomeration

More agglomerating/ harder lumps → Less agglomerating/ medium hardness lumps

Ethanol | Toluene | EtOAc | Acetone | MTBE

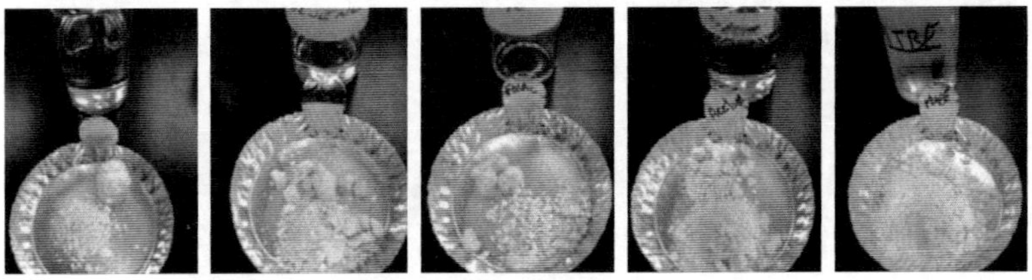

1) Ethanol
2) Toluene
3) EtOAc
4) Acetone
5) MTBE
6) THF
7) Heptane

THF → n-Heptane

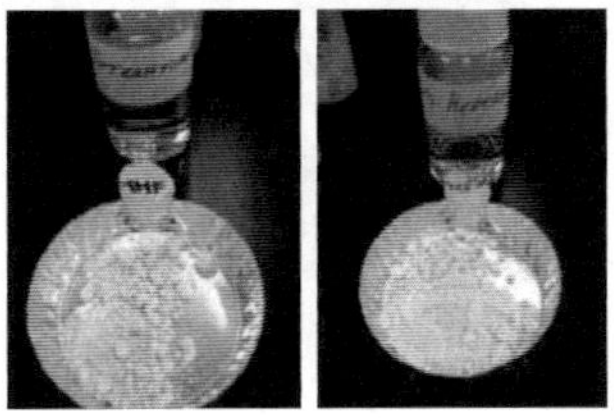

Least Agglomerating (softest lumps)

Qualitatively Heptane appears most favorable for minimizing agglomeration

NALAS Engineering Services, Inc.

Drying Involves Simultaneous Heat and Mass Transfer with Various Controlling Mechanisms

<u>Constant-Rate Period:</u> This period may be absent depending on the moisture content following filtration & blow-through. Here the cake is so wet that a continuous film of liquid exists over the solids – ie behaves as if it were bulk liquid.

$$drying\_rate = \frac{UA}{\Delta H_{vap}}(T_j - T)$$

$$t_{constrate} = \frac{m_{drycake}\Delta H_{vap}}{A}\frac{(X_1 - X_c)}{U(T_j - T)}$$

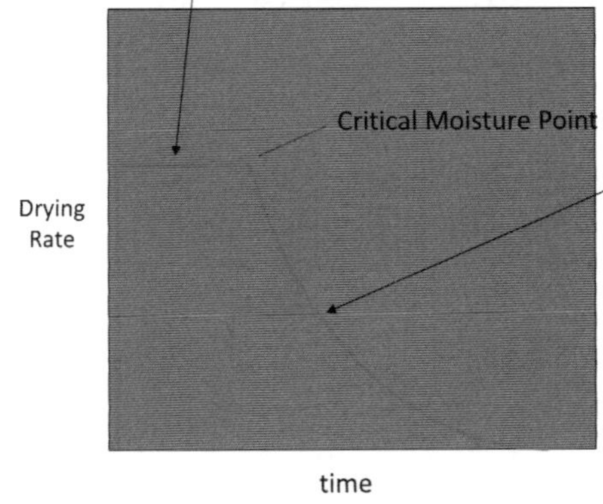

<u>Falling-Rate Period</u>: rate of moisture flow to surface Governed by internal diffusion

$$drying\_rate = aX$$

$$t_{falling\_rate} = \frac{m_{drycake}}{a}\ln\left(\frac{X_c}{X_2}\right)$$

The Critical Moisture Content(drying) varies with the thickness of the material and varies (weakly) with the rate of drying
–thus it is not a property of the material itself.
CMC should be measured experimentally.

Crystalline particles contain no interior liquid (thus drying occurs on surface of the solid) – a bed of crystalline particles can be considered a highly porous solid

NALAS Engineering Services, Inc.

DynoChem Drying Model

- Uses the Schlünder[1] model for drying
  - Models both the constant rate and falling rate of the drying process
  - Can model the effect of constant and intermittent agitation
  - Requires heat transfer properties of dryer and physical properties of solvents and solids.
    - Solvent properties are build into DynoChem, Utility available to estimate properties of solids
  - Models the drying front as it progresses from the heated wall towards the center of the cake

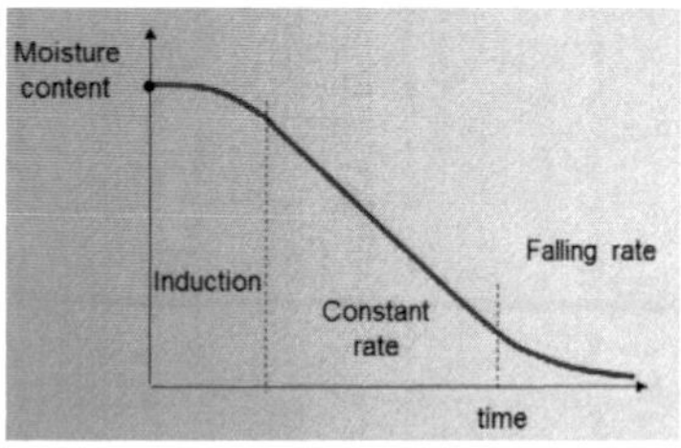

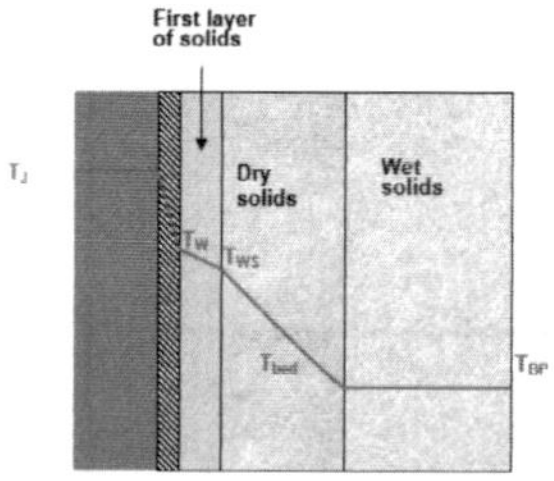

[1] Shlünder, E.U, Mollelopf, N "Vacuum Contact Drying of Free Flowing Mechanically Agitated Particulate Material" Chem. Eng. Process 18 93-111 (1984)

Large Scale AFD: DynoChem Drying Model results

Simulation of Batch Plant data

❑ Input to the model was:
- 22.50 kg of solids
- 13.88 kg EtOH
- 1.38 kg Water (vapor pressure reduced to reflect hydrate)
- Thermal conductivity of the cake consistent with 5-10 µm particles with a 2.5-3" cake depth on a 1.2 meter diameter filter with bulk density of 300 kg/m^3
- Agitate 10 min every hour at 5 rpms then continuous after 18 hours.

❑ Estimated Critical moisture content was 35%
- Moisture content, dry basis (liquid mass/solid mass)and is transition point from constant rate to falling rate, ~25% LOD

❑ Pressure was 100 mbar

❑ Jacket temperature was 45°C

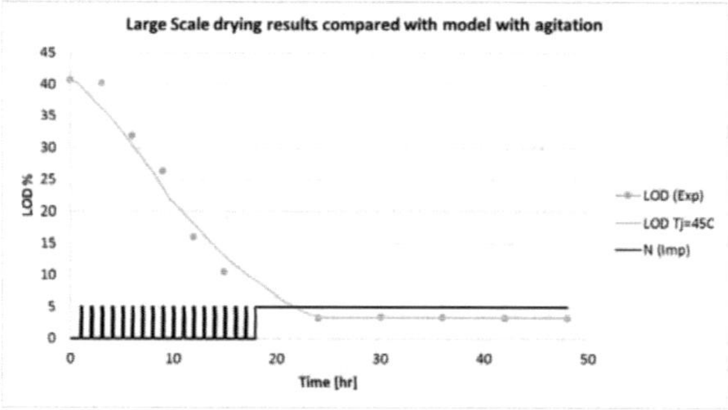

Large Scale AFD: DynoChem Drying Model

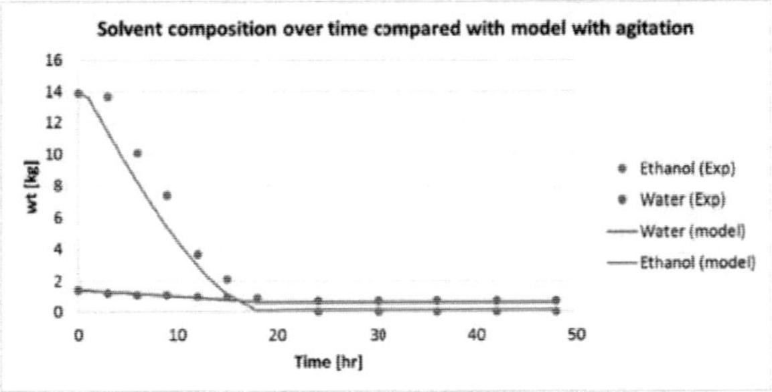

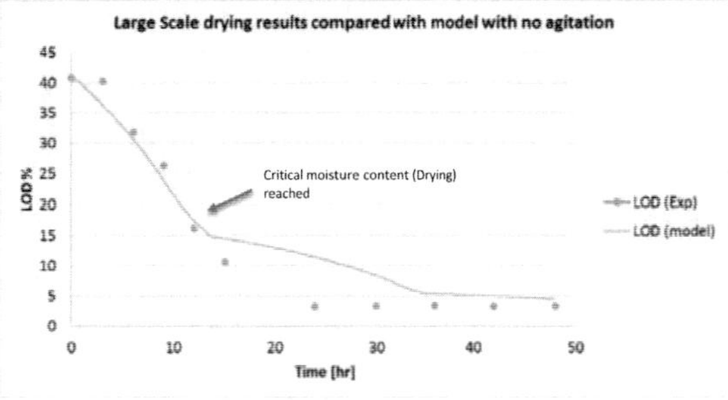

Model tracks well with not only the LOD data but also the composition of the solvent over time

Not agitating the cake does not have a major impact on drying rate in this case, until dried below the critical moisture content for drying

NALAS Engineering Services, Inc.

Parameter Sensitivity: DynoChem Drying Model

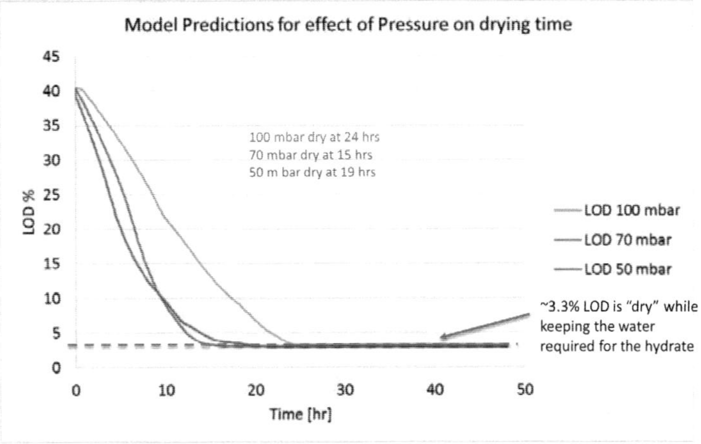

All predictions were done keeping the same agitation strategy of 10min at 5rpms every hour

NALAS Engineering Services, Inc.

Parameter Sensitivity: DynoChem Drying Model results

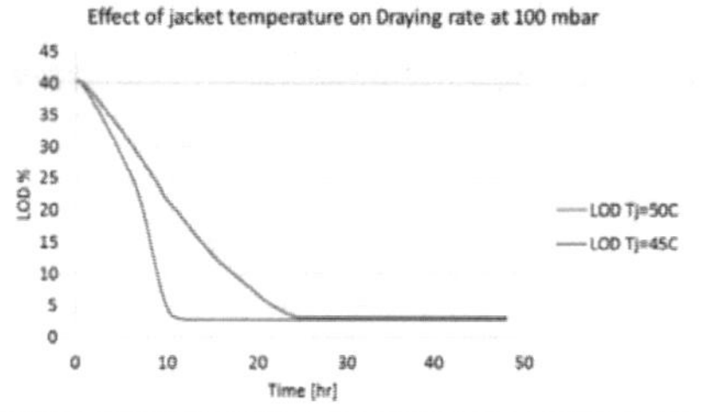

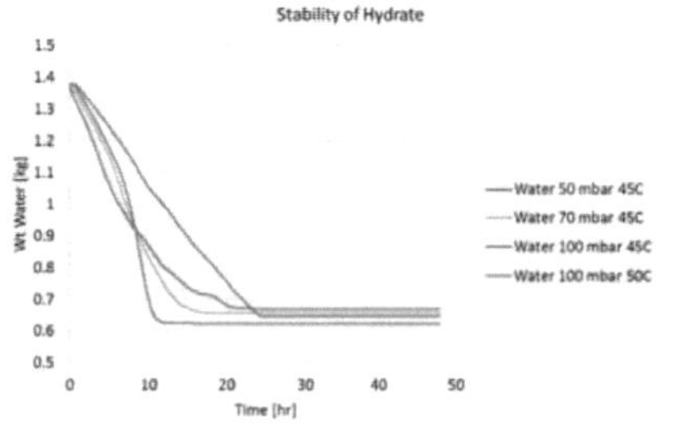

- ❑ Varying the temperature on the dryer has the most significant effect on the drying time.
  - ❑ Lower pressures, higher temperature result in faster drying rates
  - ❑ Hydrate still stable

Recommendation could be:

- ❑ No agitation until after reaching <25% LOD
- ❑ Dry at 70 mbar and 45°C

NALAS Engineering Services, Inc.

Mitigation of Agglomeration in AFD

- Modeling indicates that the Critical moisture content for drying is the same for agglomeration
  - No increase in drying rate with no agitation before critical moisture content.
  - Test if 25 % LOD is the critical moisture content
  - Test if agglomeration occurs when first dry below 25% LOD then agitate.
- Solvent screen indicated that Heptane did not tend to form agglomerates.
  - Test Filtrate rate with Heptane
  - Test for agglomeration when agitated at beginning of drying

32 NALAS Engineering Services, Inc.

Filtration Data: Ethanol at 45°C

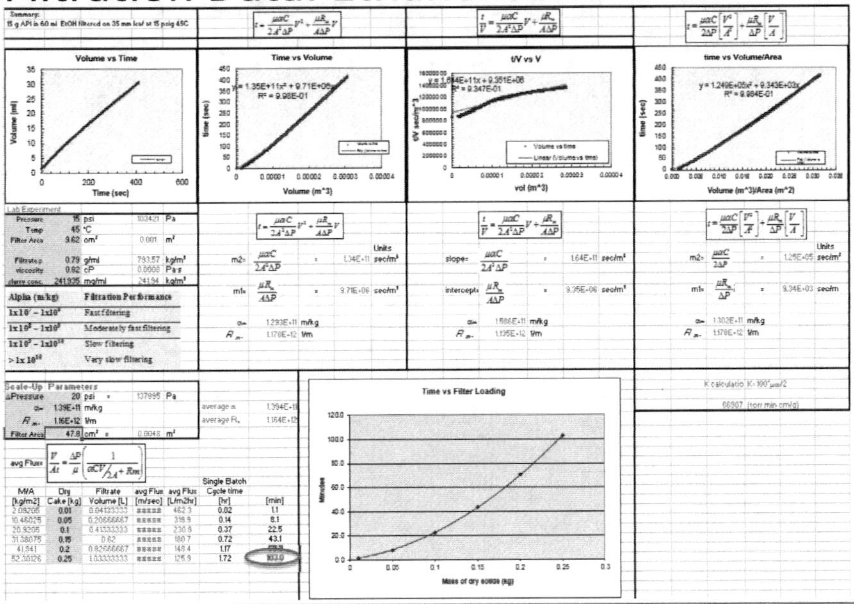

- LOD was 36% at end of filtration @ 45C
- Compared to LOD of 38% at end of filtration @20C
- Filtration rate was order of magnitude faster
  - 103 min vs 1052 min for 0.25 kg scale

Reduction in viscosity due to higher temperature reduces filtration time on scale.

Atmospheric drying to get LOD vs. Time

Drying half of the Ethanol wet cake in the hood all the way to dryness allowed for verification of the critical moisture content for drying

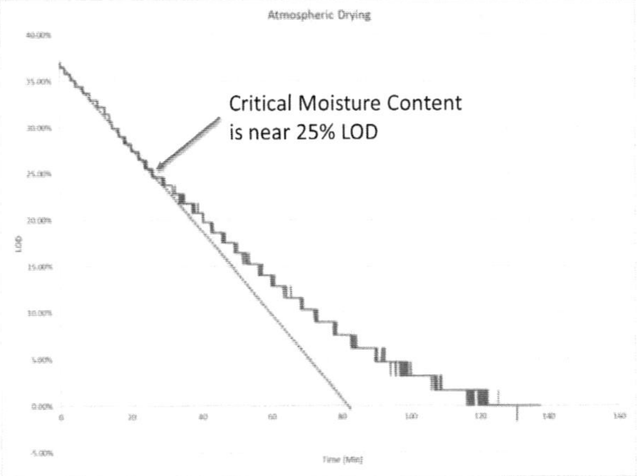

RotoVap Drying as Tumble Dryer Test: Ethanol

Ethanol wet cake dried to 23% LOD

Ethanol wet cake in flask before drying

100mBar
45C

Ethanol wet cake in flask after drying

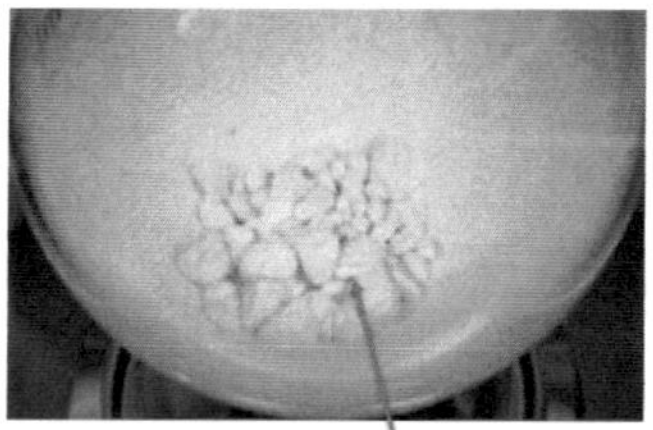

Ethanol agglomerates very Hard and make up majority of material in flask, dusting on the was is < 1mm thick

NALAS Engineering Services, Inc.

Filtration Data: Heptane at 25°C

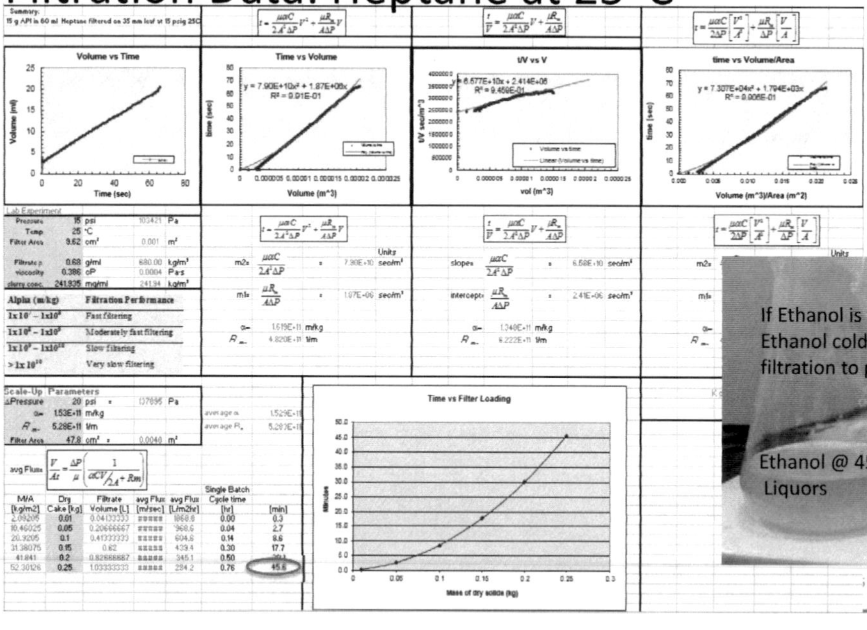

- LOD of 34% for Wet Cake from filtration
- Filtration rate 2X faster then Ethanol at 45C
  - 45.6 min vs 103 for 0.25 kg scale min

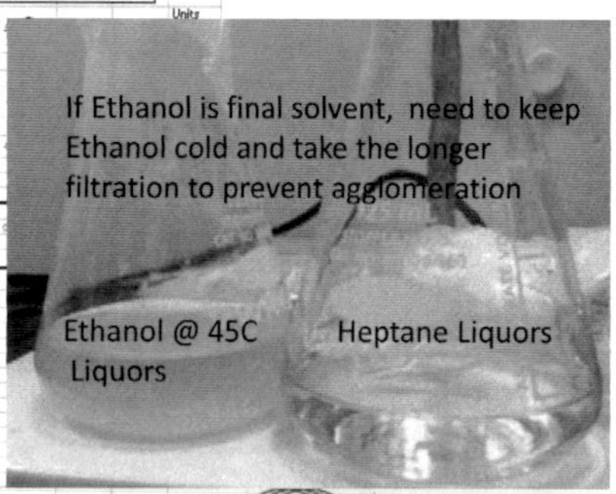

NALAS Engineering Services, Inc.

RotoVap Drying as Tumble Dryer Test: Heptane

Heptane wet cake not dried (33% LOD)

Heptan… wet cake in flask before drying … ne wet cake in flask after drying

Heptane agglomerates soft and make up small % of material in flask, layer on the was is > 20mm thick

Summary of Nalas Approach

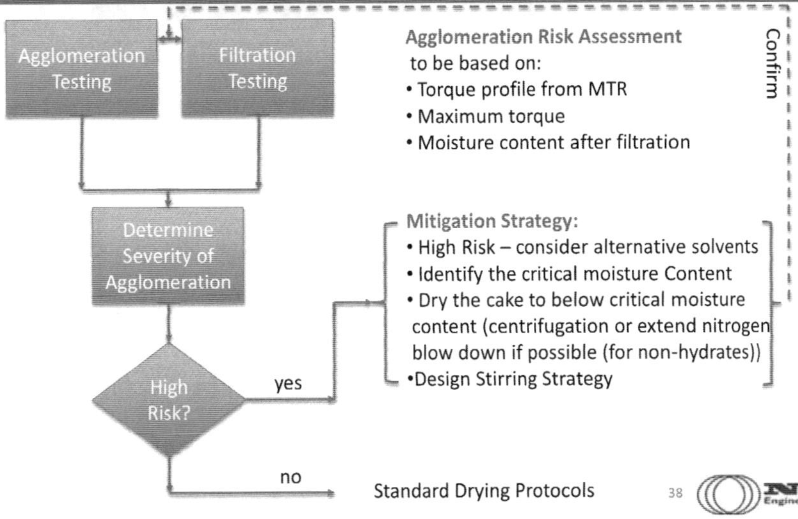

Summary:

- Optimizing filtration can help drying operation by giving a lower starting LOD for drying
- Solvent screen in the Lab RAM helps to select solvents that minimizes agglomeration.
- If solvent can not be switched out, MTR can identify what level is needed to dry to before agitation.
  - Additional stirring strategies can be investigated
    - Intermittent , continuous

39 NALAS Engineering Services, Inc.

Acknowledgments:

- Dr. Stephen Anderson
- Dr. Dave am Ende
- Brian Amato

40 NALAS Engineering Services, Inc.

Questions?

41 NALAS Engineering Services, Inc.

Mixing is Empirical

David S. Dickey
MixTech, Inc.

1

Process Development for Mixing

- Process development uses modeling and observations to understand mixing
- Scale-up "rules" are developed from test results that identify essential flow patterns and fluid mechanisms
- Careful observations at different mixing conditions provide empirical results for scale-up
- "Empirical" is not a bad word

2

MixTech®

empirical *adj.*

1. Relying upon or derived from observation or experiment.
2. Capable of proof or verification by means of observation or experiment.

The American Heritage Dictionary

- Development testing provides the needed observations
- Results can be duplicated and verified

3

MixTech®

Scale-up with Geometric Similarity

- Effective for the study of mixing
- Geometric similarity
  - tests basic flow patterns
  - studies special geometry
  - practical physical model
- Simplest form of scale-up
- Essential for solids suspension

4

MixTech®

Geometric Similarity

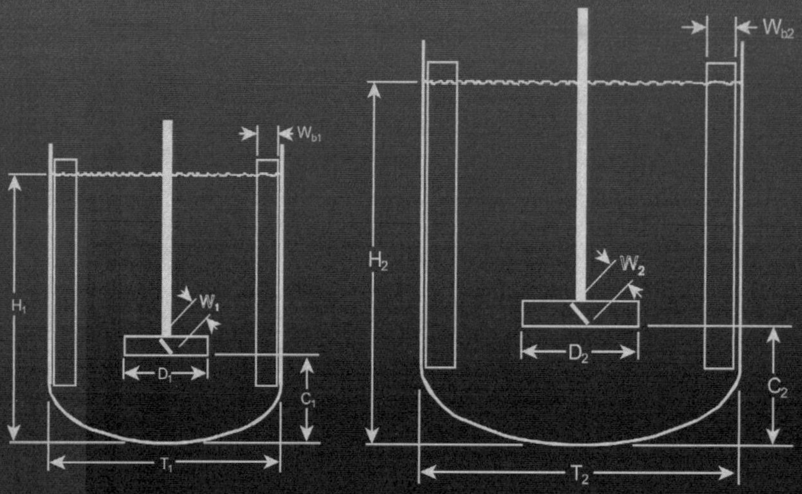

5

MixTech®

Scale-up with Geometric Similarity

- Simplified scale-up
  - one fixed geometric ratio
  - one remaining variable - rotational speed
- One mixing characteristic held constant
  - others may change
- Successful scale-up depends on keeping the correct characteristic constant

6

Scale-up of Geometrically Similar Mixers

$$N_2 = N_1 \left(\frac{D_1}{D_2} \right)^n$$

N_2 = large scale speed

N_1 = small scale speed

D_1 = small scale diameter

D_2 = large scale diameter

n = scale-up exponent

7

Simplicity and Relevance

- For the sake of simplicity
- Consider three scale-up criteria
  - equal tip speed
  - equal power per volume
  - equal torque per volume
- Using "per volume" follows scale change
- "Per volume" is identical to "Per mass" with the same fluid (density).
- Focus on turbulent conditions

8

Equal Tip Speed Scale-up

$$\pi N_2 D_2 = \pi N_1 D_1$$

$$N_2 = N_1 \left(\frac{D_1}{D_2} \right)$$

$$N_2 = N_1 \left(\frac{T_1}{T_2} \right)$$

9

Equal Tip Speed

Equal Fluid Velocity with Geometric Similarity

$$N_2 = N_1 \left(\frac{D_1}{D_2} \right)^1$$

10

Power / Volume (Turbulent)

$$Power = \mathbf{N_P}\,\rho N^3 D^5 \propto N^3 D^5$$

$$Volume = \frac{\pi}{4} T^2 H \propto T^3 \propto D^3$$

$$Power / Volume \propto \frac{N^3 D^5}{D^3} \propto N^3 D^2$$

$$\text{Scale-up:} \quad N_2^{\,3} D_2^{\,2} = N_1^{\,3} D_1^{\,2}$$

$$N_2 = N_1 \left(\frac{D_1}{D_2} \right)^{2/3}$$

11

Equal Power per Volume

$$N_2 = N_1 \left(\frac{D_1}{D_2} \right)^{2/3}$$

12

MixTech®

Torque / Volume (Turbulent)

$$Torque = \frac{Power}{Speed} = \frac{\mathbf{N_P}\,\rho\,N^3\,D^5}{N} \propto N^2\,D^5$$

$$Volume = \frac{\pi}{4}T^2\,H \propto T^3 \propto D^3$$

$$Torque\,/\,Volume \propto \frac{N^2\,D^5}{D^3} \propto N^2\,D^2$$

$$\text{Scale-up:} \quad N_2^2\,D_2^2 = N_1^2\,D_1^2$$

$$N_2 = N_1\left(\frac{D_1}{D_2}\right)$$

13

MixTech®

Equal Torque per Volume

Same as Equal Tip Speed with Geometric Similarity

$$N_2 = N_1\left(\frac{D_1}{D_2}\right)^1$$

- **Equal fluid velocity with or without geometric similarity**
- **Different from equal tip speed without geometric similarity**

14

MixTech®

Argument for Torque

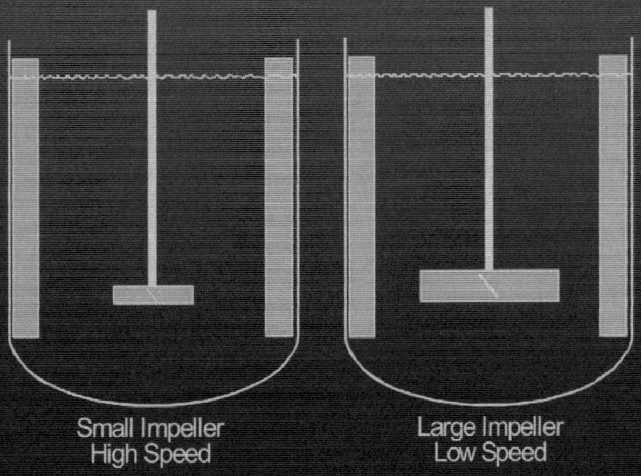

Comparable Mixing Intensity Equal Torque

15

MixTech®

Testing at Different Speeds

- Testing the effect of mixer speed is essential
- Results will change if mixing is a factor in the process
- Establish requirements for mixing
- Higher speed – more intense mixing
  - minimum speed for mixing – essential
  - maximum speed for mixing – possible or practical

16

MixTech®

Test for Mixing Effects

- Different speeds
  - +/- 5% speed = +/- 15% power
  - +/- 10% speed = +/- 33% power
- Different impellers
  - same type different diameter
    - different flow/shear ratio
    - different torque or power
  - different type same diameter
    - different flow pattern

17

MixTech®

Effects of Speed Change

- Speed change
  - tip speed increases with speed (N)
  - power increases with speed cubed (N^3)
  - torque increases with speed squared (N^2)
  - blend time decreases with speed (N)
- Difficult to distinguish effects of speed change
- Need to separate effects for good scale-up
- Better chance of separating effects with change in impeller diameter and rotational speed

18

Different Size Impellers (Same Type)

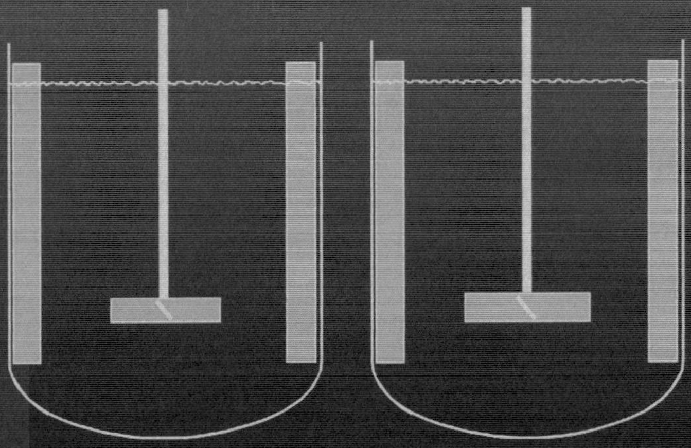

D/T = 0.33 D/T = 0.40

19

Equal Tip Speed (Same Tank)

$$\pi N_1 D_1 = \pi N_2 D_2$$

$$N_1 D_1 = N_2 D_2$$

$$N_2 = N_1 \left(\frac{D_1}{D_2} \right)^1$$

20

MixTech®

Equal Power (Turbulent) (Same Tank)

$$\rho N_1^3 D_1^5 = \rho N_2^3 D_2^5$$

$$N_1^3 D_1^5 = N_2^3 D_2^5$$

$$N_2 = N_1 \left(\frac{D_1}{D_2}\right)^{5/3}$$

21

MixTech®

Equal Torque (Turbulent) (Same Tank)

$$\rho N_1^2 D_1^5 = \rho N_2^2 D_2^5$$

$$N_1^2 D_1^5 = N_2^2 D_2^5$$

$$N_2 = N_1 \left(\frac{D_1}{D_2}\right)^{5/2}$$

22

Impeller Diameter and Speed Changes in the Same Tank

$$N_2 = N_1 \left(\frac{D_1}{D_2} \right)^n$$

- Tip Speed — $n = 1$
- Power — $n = 5/3 = 1.67$
- Torque — $n = 5/2 = 2.5$

23

Different Effects on Speed

Characteristic	D/T = 0.33	D/T = 0.40
Tip Speed	100	82.5
Power	100	72.6
Torque	100	61.8

Looking for same process results at equal measures of intensity

24

Possible Scale-up Justification

- Equal tip speed
  - shear effects
  - dispersions drop size
- Equal power / volume
  - local turbulence intensity
  - rapid chemical reaction mechanisms
- Equal torque / volume
  - liquid mixing intensity
  - fluid velocity

25

Development "Failures"

- Some failures are good
  - small failures are better than big ones
  - failures often establish design limits
  - failures may indicate critical variables
- Successful mixing development
  - may be too easy - intense mixing
  - may not be practical
- How important is mixing?

26

MixTech®

Summary

- Empirical means observation of test results
- Geometric similarity provides insight into flow patterns
- Geometric similarity simplifies scale-up by making rotational speed the variable
- Scale testing needs to identify scale-up requirements
- Test limits of successful mixing

27

MixTech®

Mixing is Empirical

David S. Dickey
MixTech, Inc.
www.mixtech.com
d.dickey@mixtech.com

28

MixTech®

29

Oil & Gas Process Separation Scale-Up Part 1: Sour Water and Sour Gas .

Engel, D., GE Water & Process Technologies

Originally presented on: 6/11/2014 15:00:00 - 15:25:00

In most Oil & Gas processing facilities, sour water and sour gas are two important streams that need to be treated for sulphur-related contaminants removal and recovery. These unwanted materials originate in the many stages of an Oil & Gas processing facility. This paper will discuss the various streams involved and how small scale equipment is related to equipment scale-up designs, and actual process efficiency increase. Real cases from past jobs will be presented.

Successful Scale-up via Simulation of Production Vessels under Lab Conditions using PAT and Thermal Analysis.

Hebrault, D., METTLER TOLEDO AutoChem

Originally presented on: 6/11/2014 15:30:00 - 15:55:00

In today's world, the chemical and pharmaceutical industries face major challenges including globalization, environmental and safety regulations, shortening product life cycle and increased time and cost pressure for commercial releases. Breakthroughs in process operations and modeling are necessary to achieve energy and material efficiency gains supporting shorter development and scale up times. Those breakthroughs are made possible in part by the lab and plant scale use of process analytical technologies (PAT) and reactor systems optimally designed for real time in situ process monitoring and accurate control of processes. For more than twenty five years the RC1e reaction calorimeter has been the golden standard for process safety and characterization studies at lab scale. Similarly, ATR-FTIR has become a mainstream PAT used by thousands of engineers and chemists around the world thanks to the success in the development of the well-known ReactIR technology. Over the past five years, METTLER TOLEDO has focused its research and development effort and investment around innovative and state-of-the-art technologies. For example,

- Real time calorimetry (RTCal),
- Expanded system platform for calorimetry investigation (volume, vessel, mixing options),
- Plug-and-play ReactIR instrument,
- Advanced chemometrics and prediction algorithm (mixing, kinetics, process safety, quantitative modeling and characterization)

They have now become the norm in R&D academic and industry labs. Several case studies will presented including for example,

- Automated control of dosing rate at lab scale according to plant cooling capabilities using RTCal and advanced predictive models
- Rapid process optimization using lab scale calorimetry, automation, and numerical simulation of thermal and mixing behavior of industrial reactors
- In-silico optimization of dosing rate at plant scale based on a kinetic model developed using quantitative ATR-FTIR results
- Qualitative use of reaction heat on a small scale to screen experimental conditions and optimization reaction conditions
- Predictive modeling and assessment of process safety profile (primary and secondary reactions) based on adiabatic and isothermal calorimetry experiments using advanced software algorithm (iC Safety).

A Continuous Diethanolamine Dehydrogenation Fixed Bed Catalyst and Reactor System.

Ringer, J., The Dow Chemical Company

Originally presented on: 6/12/2014 9:10:00 - 9:35:00

Glyphosate is currently the largest volume commodity agricultural chemical. One key reaction step in the cost-efficient route for glyphosate synthesis is the dehydrogenation of diethanolamine (DEA) to form disodiumiminodiacetic acid (DSIDA). Early process technology reportedly utilized a Raney copper catalyst in a slurry reactor. To reduce processing costs, we sought to develop a continuous fixed bed process that would provide long catalyst lifetime, high DEA conversions, and high DSIDA selectivity, eliminating the operational costs associated with catalyst attrition, catalyst recovery, and catalyst recycle using slurry reactor technology, and increasing the operational efficiency relative to slurry-based batch processing. In this summary of our development program, we highlight key findings converting from a batch to a continuous oriented process while reviewing the scale-up methodology.

CONTINUOUS PRESSURE AND VACUUM FILTRATION TECHNOLOGIES AS ALTERNATIVES TO BATCH FILTRATION

Barry A. Perlmutter
President & Managing Director
BHS-Sonthofen Inc.
Filtration Division

PRESENTATION OVERVIEW

- **Sherlock Holmes and Dr. John Watson are fictional characters of Sir Arthur Conan Doyle.**
- **Process engineers who live in the real-world can learn many things from the two of them for solving process filtration problems.**
- **This presentation will intertwine the detective techniques of Holmes and Watson with the problem solving skills required to select process filtration systems.**

NO JUMPING TO CONCLUSIONS

Discussion of the bench-top laboratory tests that are conducted for problem analysis, technology selection and scale-up.

Continuous Pressure Filtration As an Alternative to Batch Centrifuges & Batch Pressure Plate Filters

LIQUEFIED GAS – PROTEIN SLURRY

- A liquefied gas is a clean, colorless gas that is easy to liquefy. It is gaseous at normal temperature and pressure, but changes to a liquid when subjected to pressure or cooling.
- Liquefied gases are versatile process solvents.
- The most important point is to keep these gases, under pressure, so they behave as a liquid during filtration, cake washing and drying.
- The plant engineers reviewed the Rotary Pressure Filter for filtration (proteins), cake washing (lipids) and drying

Standard Laboratory Tests

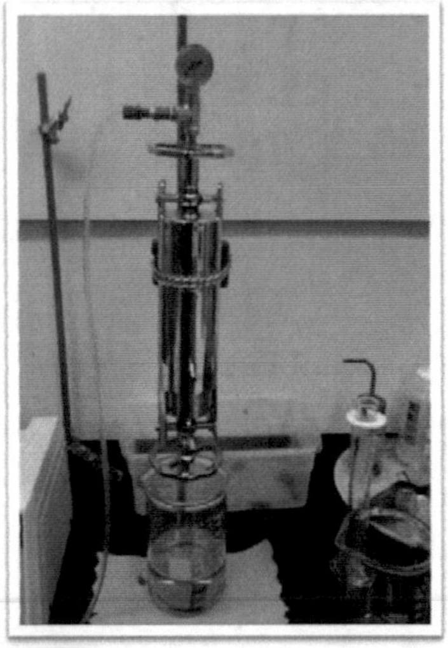

SELECTION OF FILTRATION TECHNOLOGY

CAKE STRUCTURE /PROCESS PARAMETERS FROM LAB TESTING

- **Cake Thickness and Filtration:**
- **Filter Media**
- **Cake Washing**
- **Cake Drying**
- **Cake Discharge**

BHS SONTHOFEN

Special Laboratory Tests

BHS SONTHOFEN

LABORATORY TESTING RESULTS

This test was conducted to determine the following:

- **The filter media necessary for adequate filtrate quality and wet solids discharge.**
- **The pressure necessary for adequate filtration.**
- **The solids wash strategy for acceptable product.**
- **The required filtration area for a production unit.**

BHS SONTHOFEN

BHS Rotary Pressure Filter: Operation

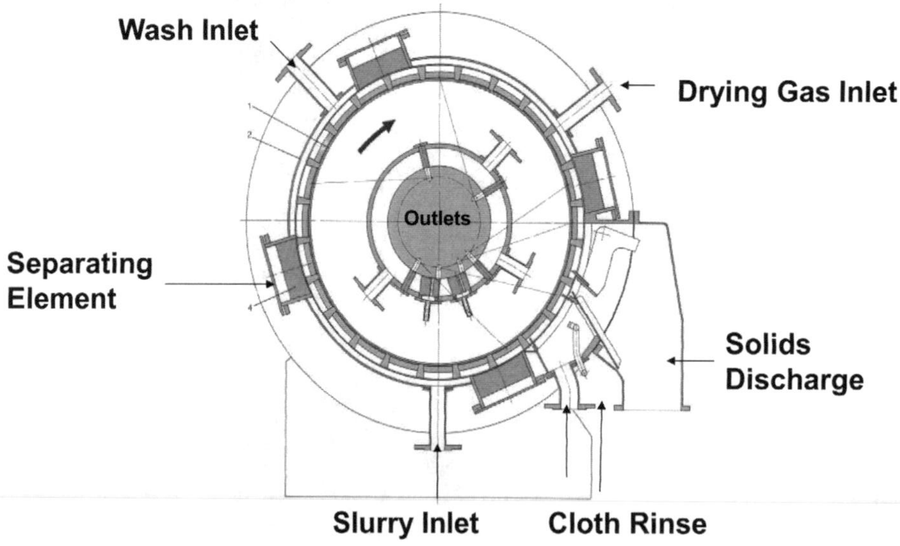

DESCRIPTION AND SPECIFICATIONS OF THE ROTARY PRESSURE FILTER

The RPF has unique features for this application to ensure that the liquefied gas remains as a liquid and to ensure the safe operation of the RPF.

- Direct-coupled drive for stable drum rotation
- Closed and pressurized housing
- Closed and pressurized bearing shields
- Secondary containment of the control head
- Pressurized control head
- Pressurized discharge housing

BHS SONTHOFEN

RPF 0.18 M² PILOT TESTWORK AND SCALE-UP

BHS SONTHOFEN

RPF 0.18 M^2 PILOT TESTWORK AND SCALE-UP

Process Parameters:
Slurry Feed Pressure
Slurry Feed Flow
Wash Pressure
Wash Flow
Dry Pressure
Drying Air Flow

RPF Parameters:
Drum Speed
Separating elements
Cake blow back
Cloth blow back
Backpressure
Cake Thickness
Filter Cloth

RPF 0.18 M^2 PILOT TESTWORK AND SCALE-UP

To evaluate the RPF performance, the site data:

Slurry solids concentration
Filtrate quantity (mother liquor, wash, blowing)
Filtrate yield
Cake Moisture
Total Cake quantity

Scale-Up From RPF 0.18 M^2 Pilot Data

- Calculate Specific Filter Performance from Pilot Testing
  = kg of dry solids/m^2/hour
- Calculate Production Area Required from Filter Performance and Client Required Production Rate
- Using the drum speed, filtration, washing and drying times and other RPF factors, the specific filter area is calculated

TYPICAL RPF 0.18 M² PILOT SCALE-UP

The scale-up is based on 224 g slurry with 1:1 composition = 112 g dry solids:

Drum revolutions:

$$n = \frac{3600\frac{s}{h}}{(4+8+15)s} \cdot \frac{270°}{360°} \cdot 0{,}85 = 85\frac{\text{revolutions}}{h}$$

270°: active angle
0.85: factor for separating elements

Specific filter performance:

$$Q = 500\frac{1}{m^2} \cdot 85\frac{1}{h} \cdot 112g = 4760\frac{kg\ dry\ solids}{m^2 \cdot h}$$

Required filter area:

$$A = \frac{20000\frac{kgdrysolids}{h}}{4760\frac{kgdrysolids}{m^2 h}} = 4{,}2\ m^2$$

Selected filter: BHS-Rotary Pressure Filter, type B16 with 5.4 m² is sufficient to operate 20,000 kg dry solids per hour.

ROTARY PRESSURE FILTER

Continuous Vacuum Filtration As an Alternative to Batch Centrifuges For Pharmaceutical Production

BHS SONTHOFEN

Standard Vacuum Laboratory Tests

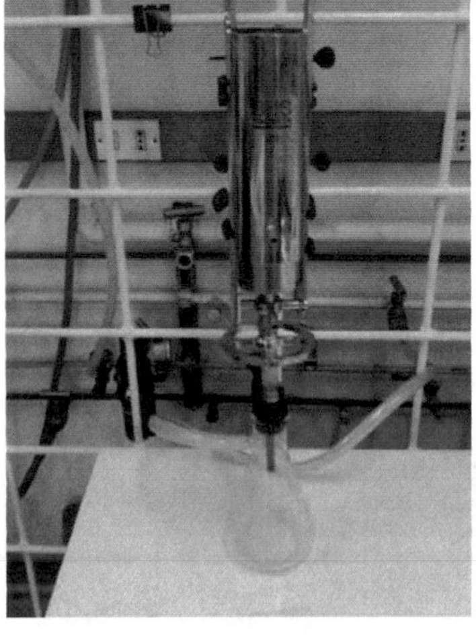

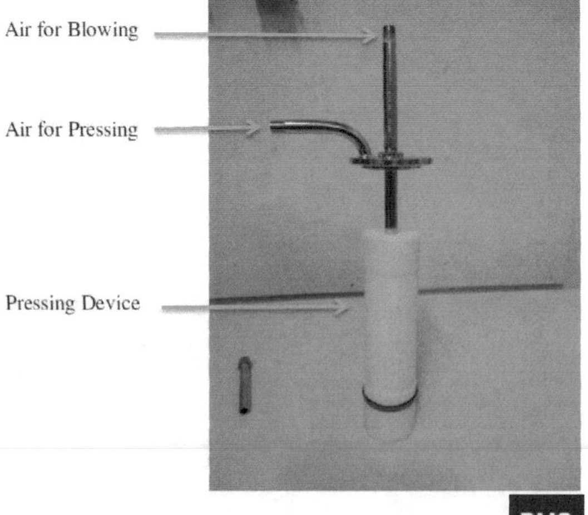

BHS SONTHOFEN

LABORATORY TESTING RESULTS

- Filtration time: 48 s extended time for sufficient washing/pressing
- Slurry Mass: 57 g
- Solids Concentration: 15 % avg.
- Filtration area: 0.002 m2
- Cake height: 8 mm
- Product flow: 71 kg/h
- Product per area: 4.28 kg/m2
- Required filtration: 0.24 m2

BHS SONTHOFEN

LABORATORY TESTING RESULTS

Trial #	Slurry Filter (sec)	Wash #1 Filter(sec)	Wash #2 Filter (sec)	Vacuum Press (mbar)
1	12	27	26	750
2	16	34	34	750
3	24	47	39	750
4	23	48	41	750

Trial	Vacuum Press (mbar)	Comp and Blow Time (sec)	Cake depth (cm)	LOD (%)
1	405	48[1]	0.7	33
2	405	48[2]	0.8	39
3	750 →405	96	0.8	29
4	560 →405	96	0.8	30

BHS SONTHOFEN

LABORATORY TESTING RESULTS

- The initial lab tests suggested a vacuum belt filter would achieve cake quality equal to or better than the current centrifuge with a major reduction in processing time.
- The footprint would be comparable to the current centrifuge and the unit would be suitable for conversion to a continuous process.
- The decision is to select a continuous-indexing vacuum belt filter for pilot testing.

Continuous-Indexing Vacuum Belt Filter

CONTINUOUS PROCESS OPERATIONS

- Vacuum Filtration
  Cake Thickness: 5 – 150 mm
- Cake Washing:
  Co-current and Counter-Current
- Drying:
  Vacuum, Convection, Pressing and Steaming

Continuous-Indexing Vacuum Belt Filter

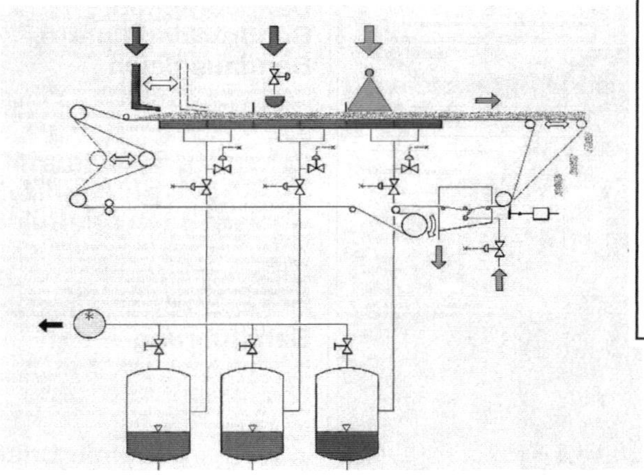

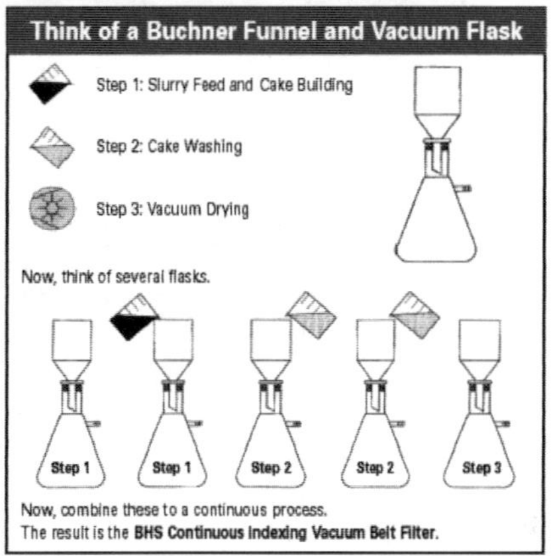

BHS SONTHOFEN

Pilot Continuous-Indexing Vacuum Belt Filter (0.1 m2)

BHS SONTHOFEN

Pilot Continuous-Indexing Vacuum Belt Filter (0.6 m2)

Pilot Continuous-Indexing Vacuum Belt Filter Scale-Up from 0.6 m2

Process Step	Number of Zones (0.5 m width)
Filtration	1
Wash 1	1
Wash 2	1
Drying	3 (2 x Pressing + 1 x Vacuum)
Total Length	3 meters

Pilot Continuous-Indexing Vacuum Belt Filter Cleaning Tests On 0.6 m2

Sample ID	Description	ug/cm2
1	Slurry inlet tray	Non-detectable
2	Wet cake discharge	0.08
3	Belt wash box	0.04
4	Belt Filter Tray	Non-detectable

Swab Cleaning Test
1. Filter rinsed and all visible material removed
2. Filter sprayed with isopropyl alcohol
3. Apply solution of 10% sodium thiosulfate for 30 minutes
4. Filter rinsed
5. Swab Testing

BHS SONTHOFEN

Pharma Pilot Continuous-Indexing Vacuum Belt Filter

BHS SONTHOFEN

SUMMARY "LEARNING TO TELL THE CRUCIAL FROM THE INCIDENTAL."

- **Engineers must evaluate all outcomes to make an informed and successful decision.**
- **Technical evaluation, laboratory & pilot testing are critical for successful decisions and projects.**

SUMMARY "OCCASIONAL SILENCE" & "EMPLOYING DISTANCING

- **The take-away is that close collaboration between the client and the vendor provides for creative problem-solving and process filtration solutions to achieve the production objectives.**

THANK YOU!

Barry A. Perlmutter
President & Managing Director
BHS-Sonthofen Inc.
14300 South Lakes Drive
Charlotte, North Carolina 28273-6794
Tel: 704.845.1190; Fax: 704.845.1902
E-mail: barry.perlmutter@bhs-filtration.com

Advantages and Limitations of a Nitration Reaction Performed in Batch vs. Continuous Mode.

Fanfair, D., Bristol-Myers Squibb

Originally presented on: 6/12/2014 10:10:00 - 10:35:00

Nitration reactions are commonly utilized in the pharmaceutical industry to synthesize aromatic amines. Developing safe, scalable nitrations can be challenging since these reactions are generally highly exothermic and result in the production of energetic, nitro containing compounds. While incorporating a continuous configuration for these reactions may help address some of the safety concerns due to a reduced in-process volume of reactive species and greater heat transfer, batch configurations may also offer significant advantages. This presentation will provide a comparative case study on the advantages and limitations of continuous and batch configurations for the nitration of a pharmaceutically relevant pyrrole. Included will be details of the laboratory scale process development activities and process demonstrations conducted at scales from 500g to 2.5 kg.

Scale Up of Bio-Succinic Acid: Myriant's Technical Approach

Process Development Symposium – AIChE
Philadelphia, PA
June 10-12, 2014

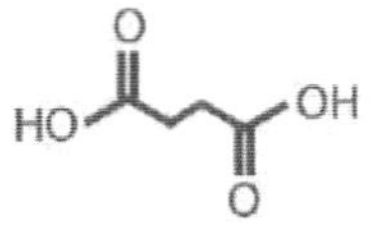

Rajesh (Raj) Dasari, Ph.D.
Sr. Process Engineer
Process Development Group
R&D Center, Woburn, MA
Myriant Corporation

Why Bio-Chemicals ?

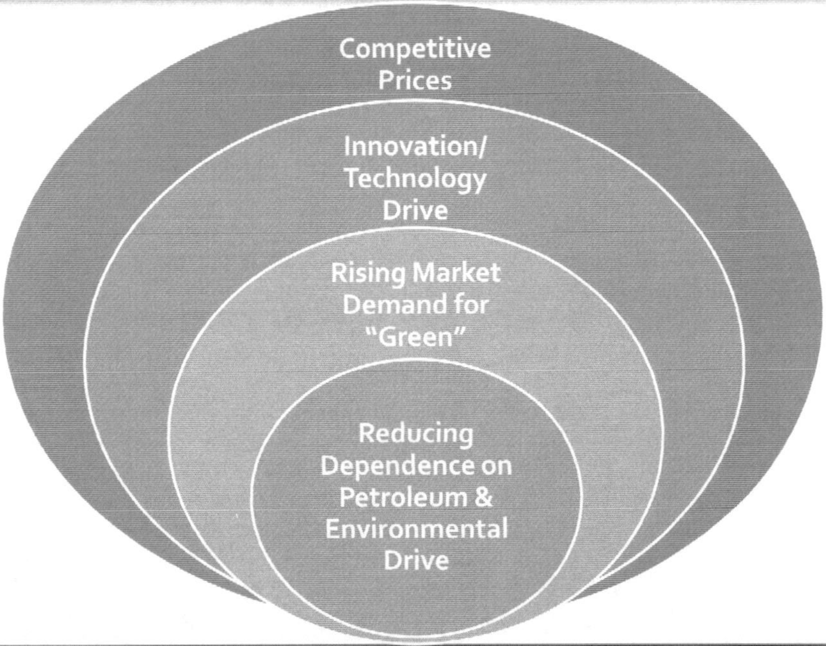

Commodity vs. Fine Chemicals

Myriant

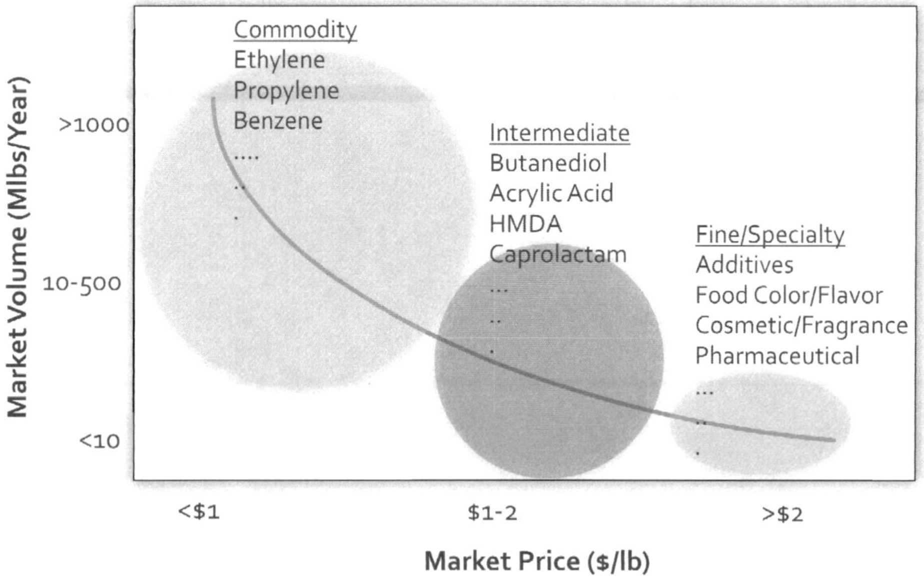

Myriant's Footprint

Myriant

The Company

- Formed in 2009.
- HQ , R&D in Greater Boston area.

- Bio-Succinic Acid commercial plant in Lake Providence, LA.
- 105 people organization (R&D, engineering, manufacturing, sales & marketing).

Roadmap for Biochemical Commercialization

Myriant

Lactic Acid
Succinic Acid
Muconic Acid

30 million lbs / Year Commercial Plant in Lake Providence, LA

Strong Market Demand Supporting US Expansion Plant

Sustainable and Cost-Advantage Biochemicals Platform

Chemistry Refined...Naturally 5

Catalysis: Step Up to Higher Value Chemicals

Myriant

Primary Building Block → ***Chemical Catalysis*** → ***Secondary Building Block*** → ***Consumer Products***

Myriant Product Pipeline

- Succinic Acid **$7.5B**
- Lactic Acid **$15B**
- Muconic Acid **$22B**

Current Succinic Acid Applications
Butanediol
Replacements for Adipic & Phthalic Anhydride
PLA
Acrylic Acid
Adipic
HMDA
Caprolactam
HDO
CHDM

Chemistry Refined...Naturally 6

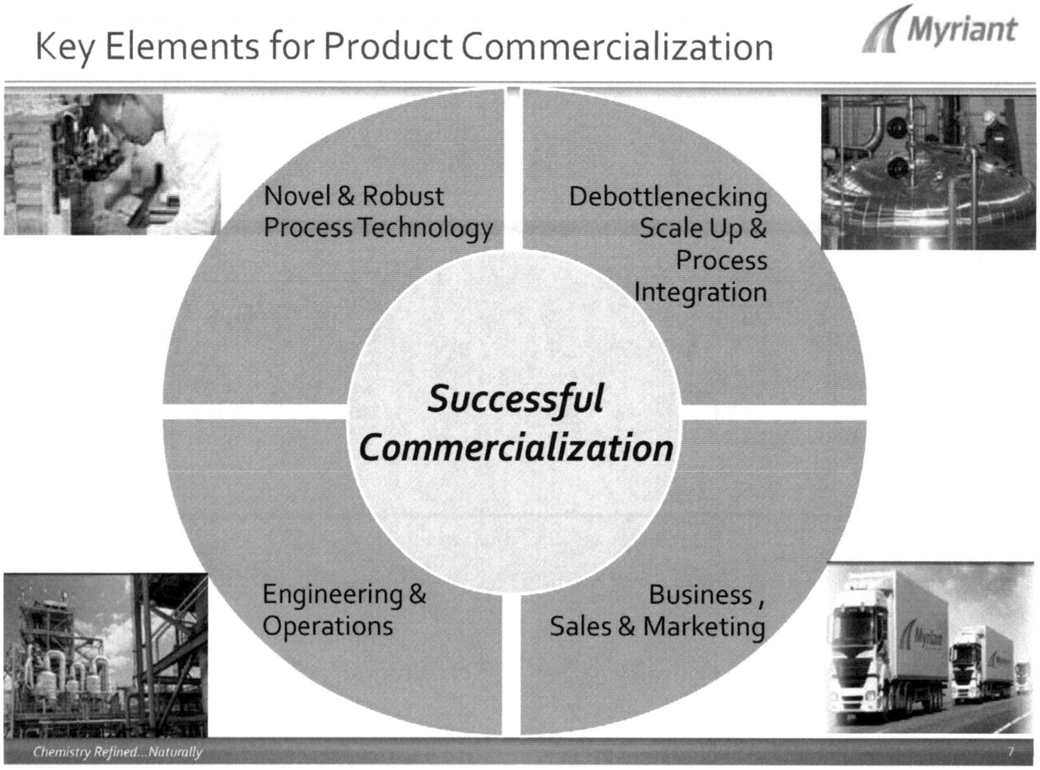
Key Elements for Product Commercialization
Myriant
Novel & Robust Process Technology
Debottlenecking Scale Up & Process Integration
Successful Commercialization
Engineering & Operations
Business , Sales & Marketing
Chemistry Refined...Naturally
7

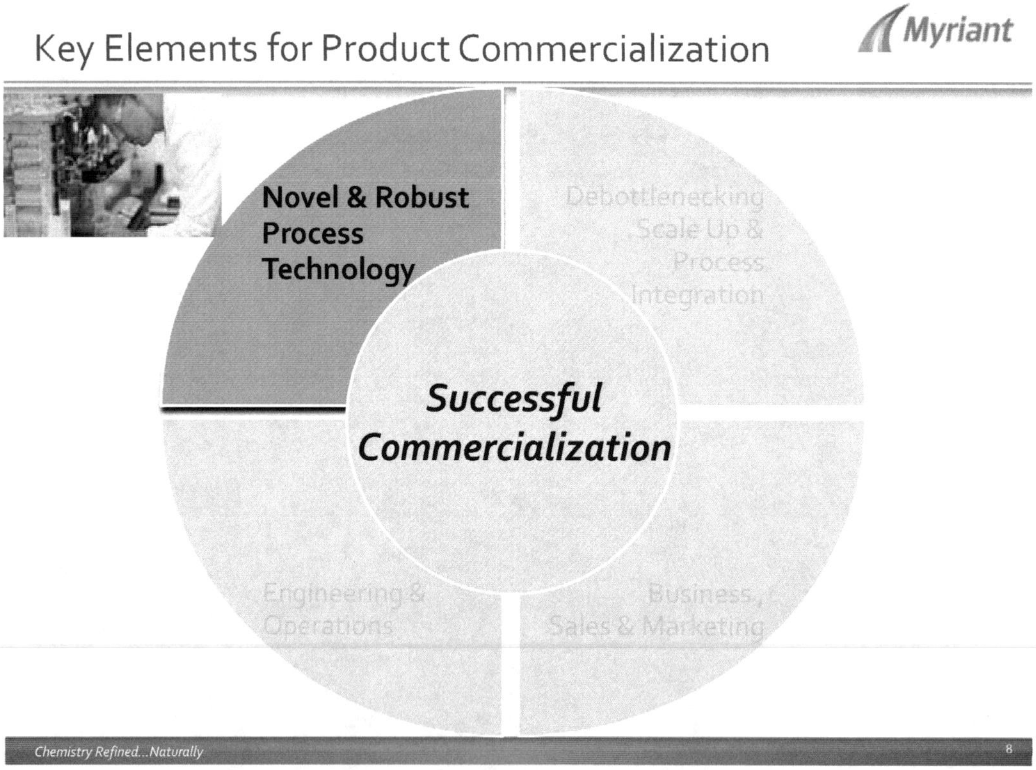
Key Elements for Product Commercialization
Myriant
Novel & Robust Process Technology
Debottlenecking Scale Up & Process Integration
Successful Commercialization
Engineering & Operations
Business , Sales & Marketing
Chemistry Refined...Naturally
8

Organism Development - Building Blocks for the Pathway to Numerous Biochemicals

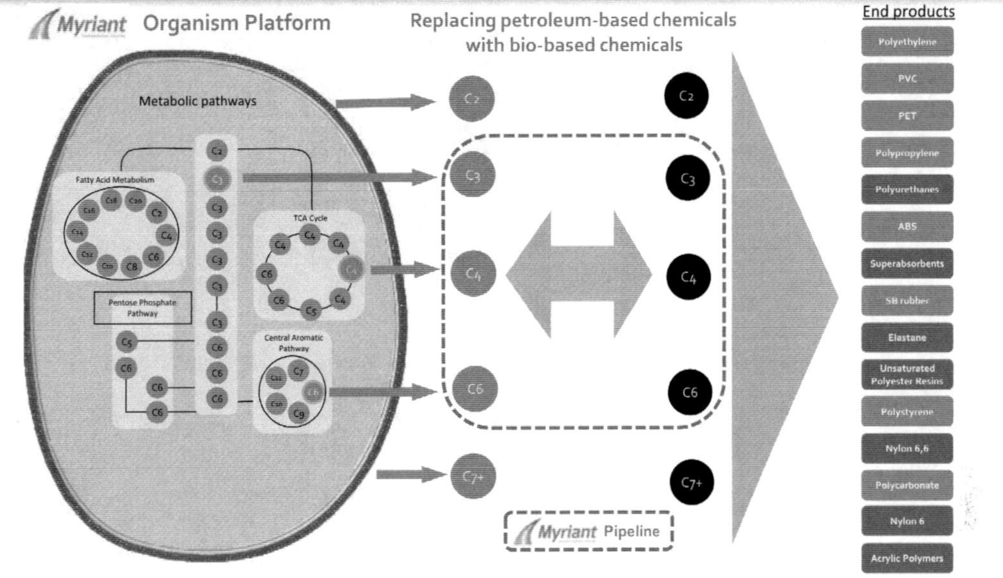

90% of all petrochemicals can be produced biologically through an organism.

Organisms-Key to Strong Technology Platform

Desired Organism Attributes

- High titer and high yield
- High productivity
- Minimal media requirement
- Bio level safety, regulatory (non-GMO)
- Robustness

Goals

- Reduce production and capital costs
- Reduce need for costly purification
- Scalable process

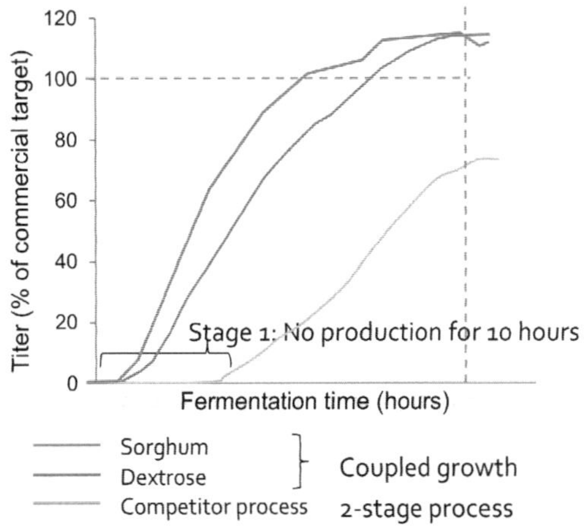

Producing product even in growth phase.
In some competitor cases CO_2 or only cell mass or by products.

Downstream Product Recovery & Purification

Myriant

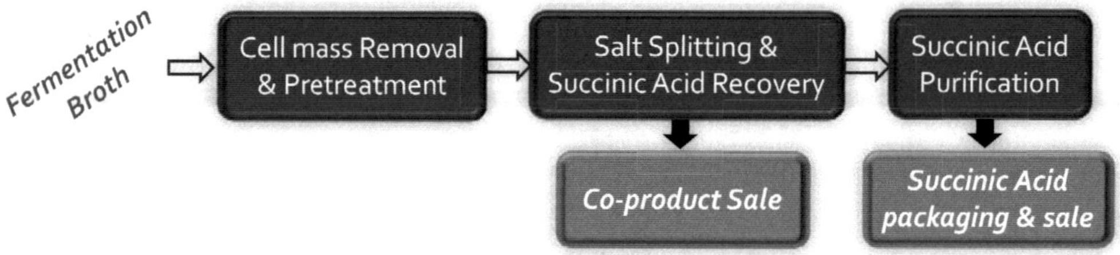

- Downstream is sufficiently robust to maintain process sustainability allowing improvements and variations in a biological process at large scale.
- The process is capable to produce multiple grade products to cater to different market needs.
- Continued process improvements, intensification and simplification towards lower OpEx & CapEx.

Chemistry Refined...Naturally 11

Key Elements for Product Commercialization

Myriant

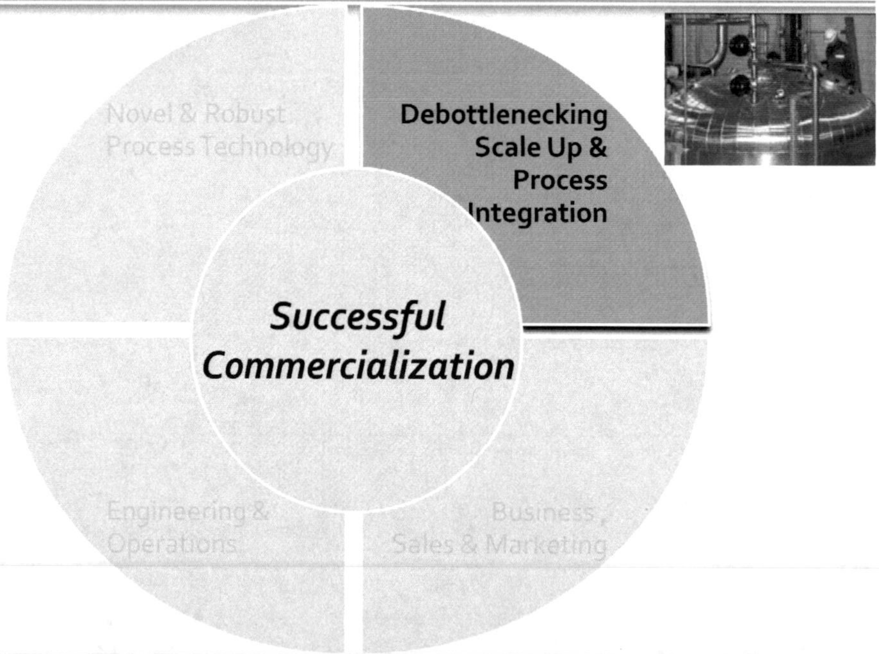

Chemistry Refined...Naturally 12

Technology Scale Up – Piloting

Myriant

Own Pilot Plant		***Contract Pilot Plant***	
CapEx required	(negative)	Minimal to No CapEx required	(positive)
Longer lead time for operations start up	(negative)	Faster track to commercialization	(positive)
Remains Company's Asset	(positive)	Not Company's Asset	(negative)
Lower OpEx	(positive)	Higher OpEx due to tolling prices	(negative)

Technology Scale Up – Piloting

Myriant

Raw Materials → **Conversion Process** → *Products & By-products*

Utilities → **Conversion Process** → *Waste*

- Material balance
  - Provides opportunity to identify critical areas for yield loss reduction.
- Energy balance and utility estimation
  - Non-trivial expense. Need critical evaluation of this aspect and engineering solutions to reduce utility usage.
- CapEx & OpEx estimation
  - Early stage cost model development is vital and provides opportunity to identify viability and critical areas for process improvements.
- Basic engineering:
  - Utilize the above listed activities to generate data for basic engineering of plant.

Technology Scale Up – Debottlenecking

Myriant

Requires the right team – Internal team combined with external engineering partnership.

Study, assess the impact, and address key commercialization risk elements.

- Energy audit by pinch analysis
- Recycle loops
- MOC studies
- Water usage
- Waste water load
- Unintentional material losses (equipment hold up volumes / vessel heel volumes, pre-steady state losses, etc.)
- Value engineering

Communications with equipment/unit operations vendors is vital.

Apply lessons learned and knowledge to commercial plant engineering.

Chemistry Refined...Naturally 15

Key Elements for Product Commercialization

Myriant

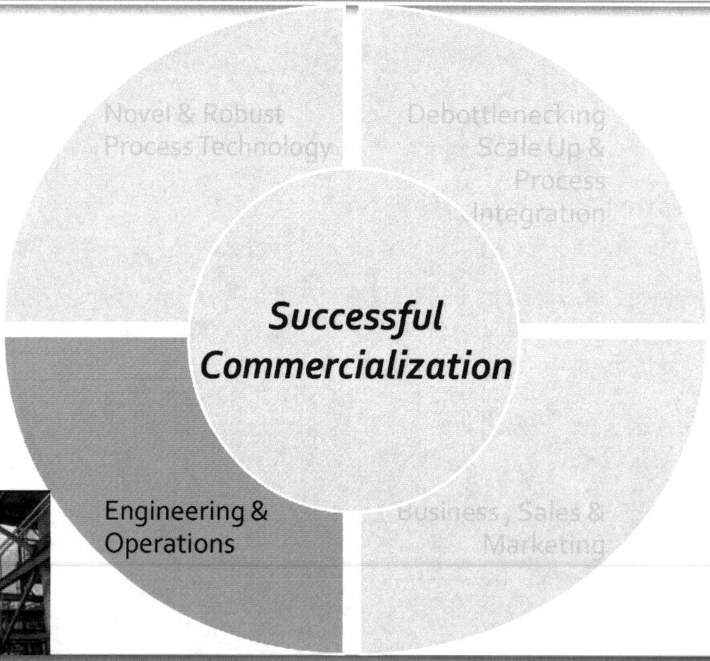

Chemistry Refined...Naturally 16

Engineering & Operations

This is another critical functional team to drive the engineering aspects of scale up from pilot to commercial scale.

RFP, Bids evaluation, Vendor Selection
Engineering, Procurement & Construction

↓

Commissioning and start up.

↓

Product ramp up to design capacity
Steady state operations.

Succinic Acid Production Phase 1: 30 Million Lbs. Lake Providence Plant

- **First** Bio-Succinic Acid plant in U.S.
- **30 million lbs / Year** Commercial Plant in Lake Providence, Louisiana
- Production is on going.

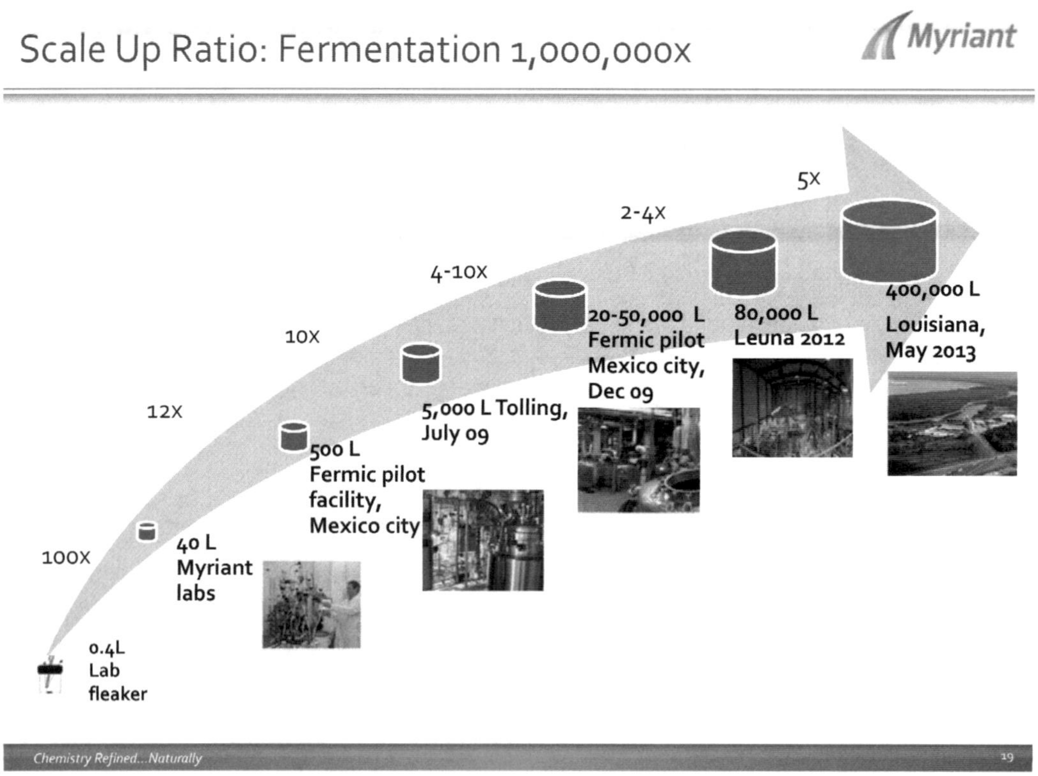
Scale Up Ratio: Fermentation 1,000,000x
Myriant
100X
12X
10X
4-10X
2-4X
5X
0.4L
Lab
fleaker
40 L
Myriant
labs
500 L
Fermic pilot
facility,
Mexico city
5,000 L Tolling,
July 09
20-50,000 L
Fermic pilot
Mexico city,
Dec 09
80,000 L
Leuna 2012
400,000 L
Louisiana,
May 2013
Chemistry Refined...Naturally
19

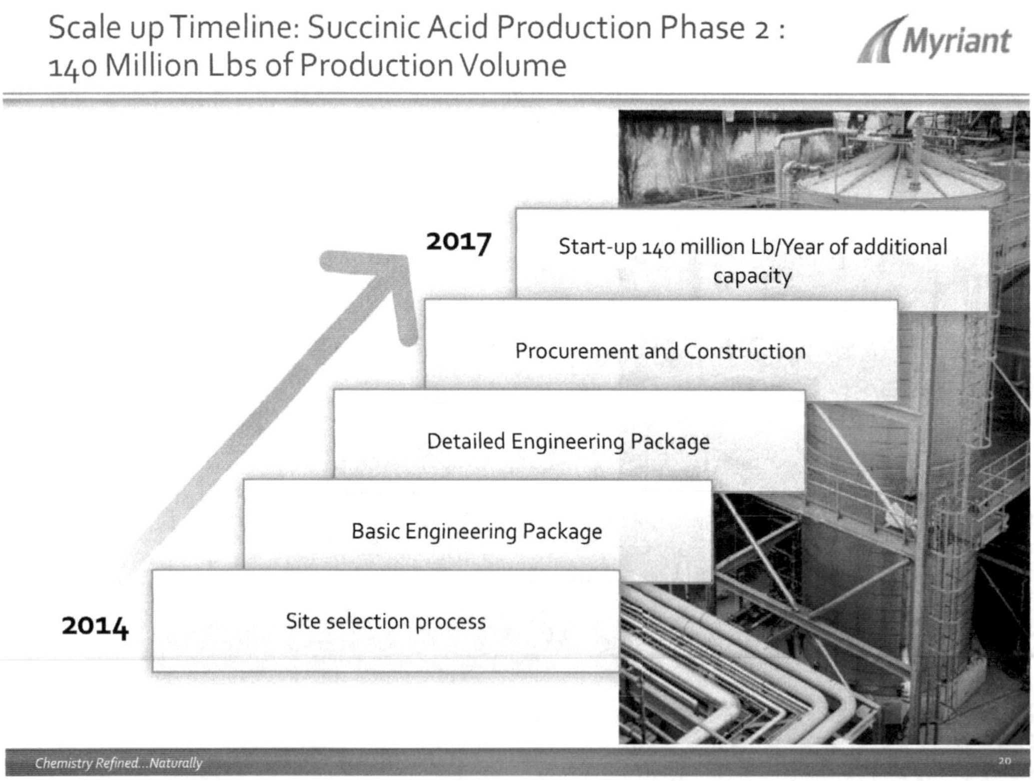
Scale up Timeline: Succinic Acid Production Phase 2 :
140 Million Lbs of Production Volume
Myriant
2017
Start-up 140 million Lb/Year of additional capacity
Procurement and Construction
Detailed Engineering Package
Basic Engineering Package
2014
Site selection process
Chemistry Refined...Naturally
20

Key Elements for Product Commercialization

Myriant

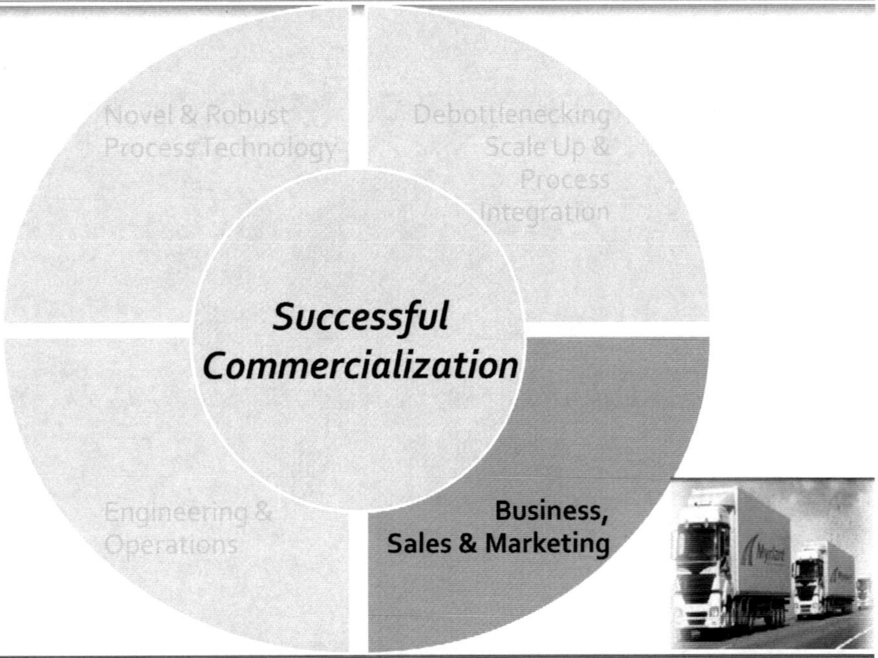

Business, Marketing & Sales

Myriant

Prepared and implemented the;

Market penetration and growth strategy.

Product grades, sales and distribution strategy.

Create partnerships & collaborations. ❖ PTT GC ❖ ThyssenKrupp Uhde ❖ Davy Process Tech. ❖ Sojitz ❖ Bayegan	Bio-SAC Product branding & marketing. ❖ Myrifilm® ❖ Myrithane™ ❖ Myriflex ™	Off-taker agreements for product sales and support generating revenues.

Product Quality is Paramount: Tested and Verified by Customers

Myriant

Bio-Succinic Acid product quality testing

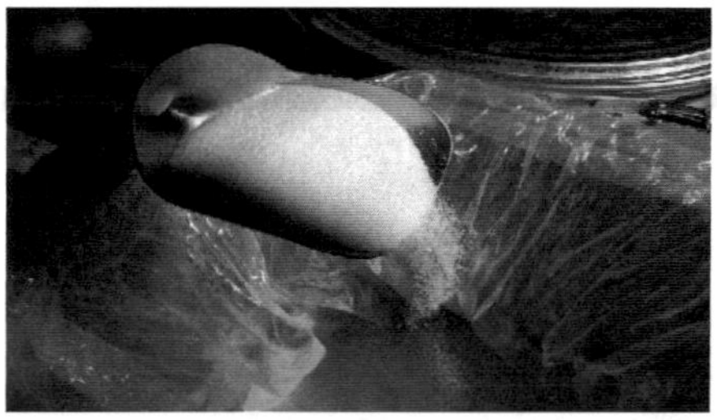

Customer Application
Polymers
Coatings
Plasticizers
Urethanes
Solvents

3rd Party Independent Labs

CAL POLY SAN LUIS OBISPO	Pennsylvania College of Technology PENNSTATE	1ST SOURCE RESEARCH	Polymer Diagnostics Inc
COATINGS	BIOPOLYMERS	URETHANES	PLASTICIZERS

Summary: Project Road Map - Stage Gates

Myriant

Business Opportunity	Process Options & Conceptual Design	Design & Engineering	Project Execution	Commission / Start up
• Strategy • Market • Customers • Product Needs • Key Business risks	• Process Options • IP Landscape • Target Costs • Process Development • Technical Risks • Resource Plan • Pilot scale tests • Basic Engineering	• Finalize scope • Value Engineering • Plant Design • RFP, bids evaluations, vendors selection and contracts.	• Secure permits • EHS management • Procurement and construction.	• Vendor support for commissioning • Operator training program • Control & monitoring tools in place. • Production ramp up. • Steady state operations
Lead: **Business**	**R&D**	**Eng.**	**Eng.**	**Operations**

Leveraging Succinic process development and commercialization experience to accelerate development of other pipeline products.

Priorities going forward

Myriant

- Continue to serve our customers in Americas, Europe and Asia with high quality product supplied from Lake Providence.
- Launch more Succinic Acid derivative products.

25

Myriant

Thank You!

Renmatix™

Scaling up the Plantrose™ Process

Dan Floyd, Process Technology Director

AiCHE, June 2014

Renmatix™

BIOMASS TO SUGAR.
FASTER & CHEAPER. REALLY.

Renmatix

1

FASTER: Plantrose™ Process - Fast & Flexible

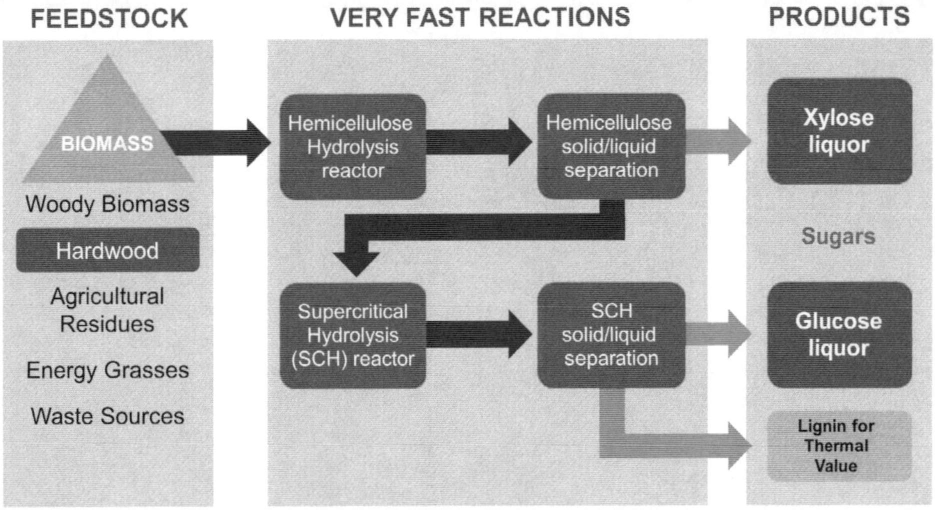

ADDRESSABLE MARKET INCLUDES A WIDE VARIETY OF BIOMASS TYPES & DOWNSTREAM USES FOR SUGAR

Renmatix 2

CHEAPER: Lowest Cost Pathway to Cellulosic Sugar

Disruptive Technology = **Disruptive Economics**

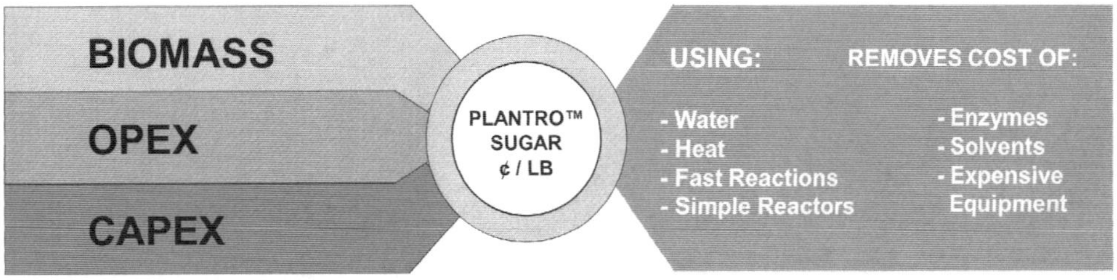

1st Facility: Competitive with Corn & Cane

Renmatix 3

REALLY: Strategic Partners Investment & JDAs

BASF The Chemical Company

WM WASTE MANAGEMENT

UPM

KEY INFLUENCER RECOGNITION

Renmatix

4

Piloting Purpose – Enabling Project Execution Succe$$

DEMONSTRATION OF PROCESS AND PRODUCTS

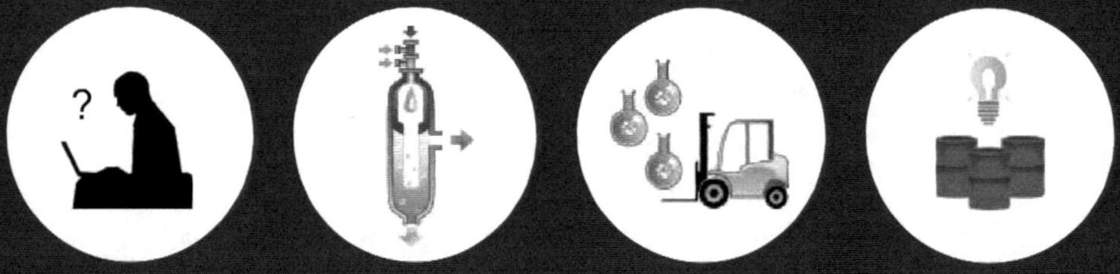

Testing process on lower cost and risk facility	Understanding and addressing process safety concerns	Evaluating chemistry and mechanical reliability that achieve your economic targets	Produce product for customer offtake agreements

Renmatix

5

Piloting Purpose – Enabling Project Execution Succe$$

DEMONSTRATION OF PROCESS AND PRODUCTS

A + B = C

- Defining the normal operating envelope
- Develop start up / shut down strategies
- Train organization on developing Process Technology and Operating Discipline
- Develop process control code, quality control, and best practices for reliable operation

Renmatix

6

Piloting Purpose – Enabling Project Execution Succe$$

DEMONSTRATION OF PROCESS AND PRODUCTS

Ready, Aim, FIRE!

What is the RIGHT scale for piloting / demonstration

Doing too many demonstrations at once

What are you really trying to get good at?

Trying to meet your CAPEX targets in your pilot plant

Not maximizing your pilot asset's output

Renmatix

7

Things That Will Help

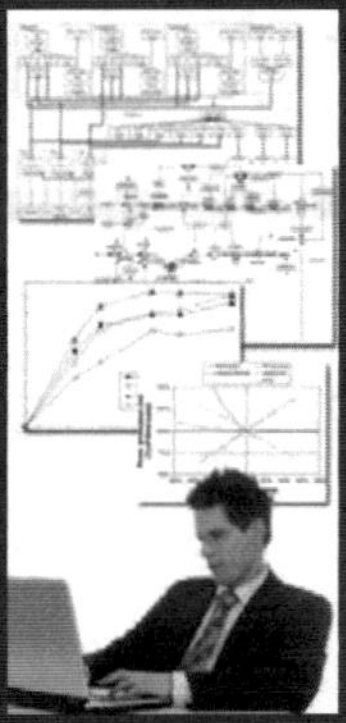

Data, data, data and the statistics you must apply

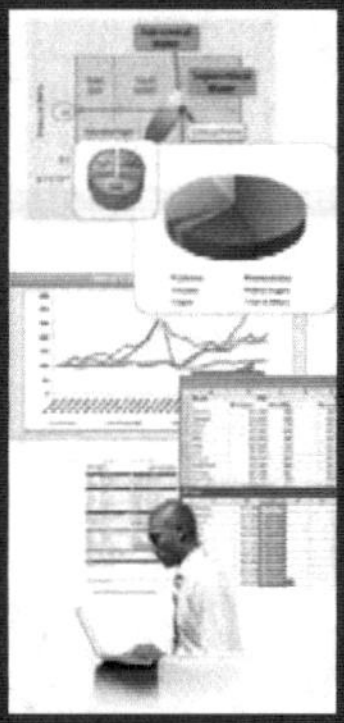

Models and more models

(Dynamic process, CFD, combined ASPEN / economic)

Great process chemists and the analytical chemists they need

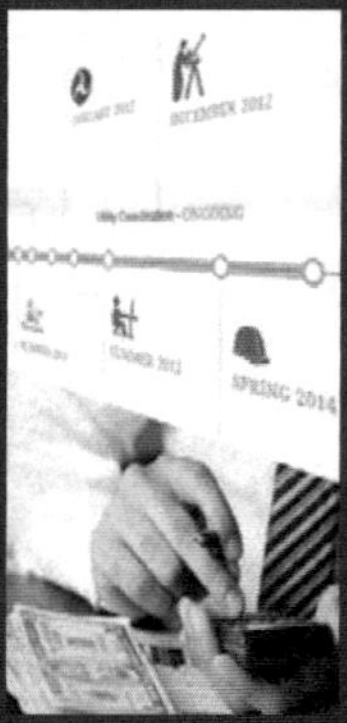

Knowing what will have the greatest $$ or timeline impact on project execution

A Great Team!

Renmatix

8

WHERE WE ARE TODAY:
Georgia - Robust Demonstration Facility

Processing hardwood biomass into sugars 24/7

Creating the enabling intermediates for partners' renewable chemical & fuel production

Plant capacity: 3BDT/day

Renmatix

9

WHERE WE ARE TODAY:
Pennsylvania – World Class Analytical & Technology Center

World Class Tech-Center Opened 2013

- Exploring new types of biomass with state of the art analytical capabilities -

- Extend our collaboration (JDAs) with downstream customers -

- Integrate further process enhancements -

Renmatix

10

NEXT CHAPTER

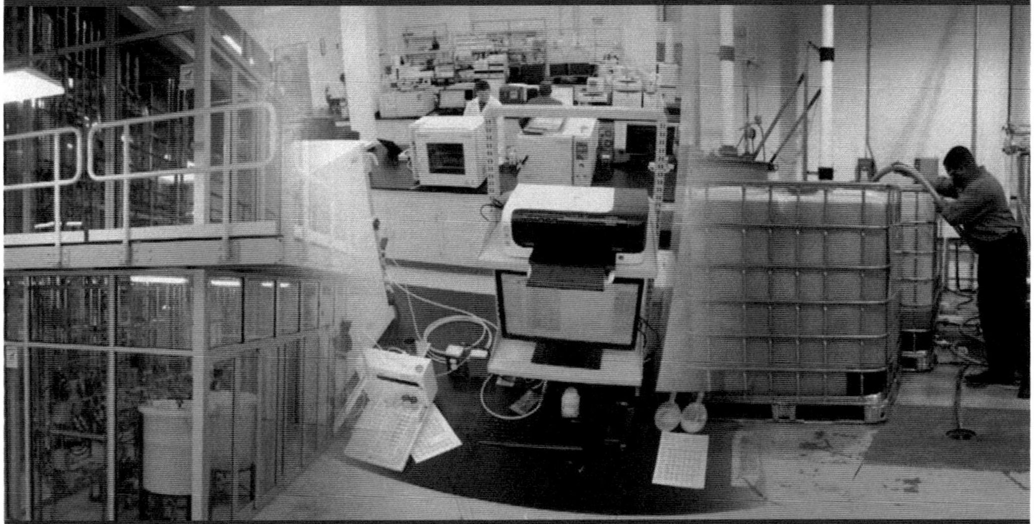

MILESTONES TO FIRST COMMERCIAL FACILITY

Renmatix

11

Renmatix™

BUILDING BLOCKS FOR BIOCHEMICALS

Renmatix

Wood to Jet fuel -
– Sustainable Energy Pathway (SEP) Analysis

Science, Engineering and Education for Sustainability (SEES)

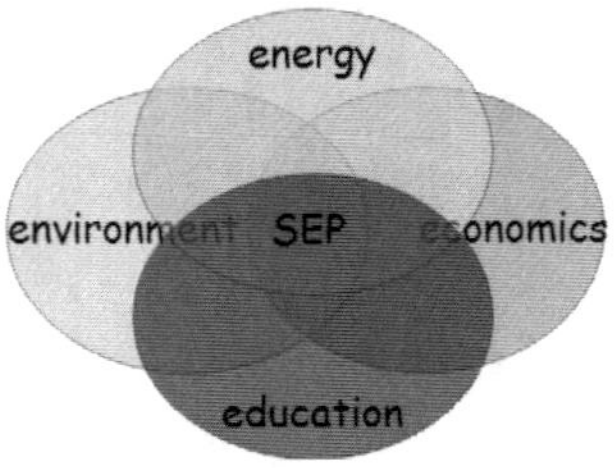

Hemant Pendse,
Director FBRI UMaine
pendse@maine.edu

June 12, 2014

Acknowledgements

- DOE EPSCoR project (DE-FG07ER46373)from 2007-2015
- DLA award on wood to jet fuel from 2011-2013
  - Sponsored by the Logistics Research and Development Program at the Headquarters of the Defense Logistics Agency, Fort Belvoir, VA. (SP4701-11)
- NSF SEP award (1230908) from 2012-1016

Aviation Fuel Readiness Level (FRL)								
Scale 1-9	fuel certificat'n	fuel production						
Feedstock Readiness Level (FSRL)								
Scale 1-9		feedstock production	feedstock market	feedstock policy, reg, social	feedstock linkage to conversion			
Energy Pathway Readiness Level (EPRL)								
Scale 1-10	sustain-ability certificat'n	pathway production	pathway markets	pathway policy, reg, social	value chain integration	geographic infra-structure	regional economic drivers	external markets, policy

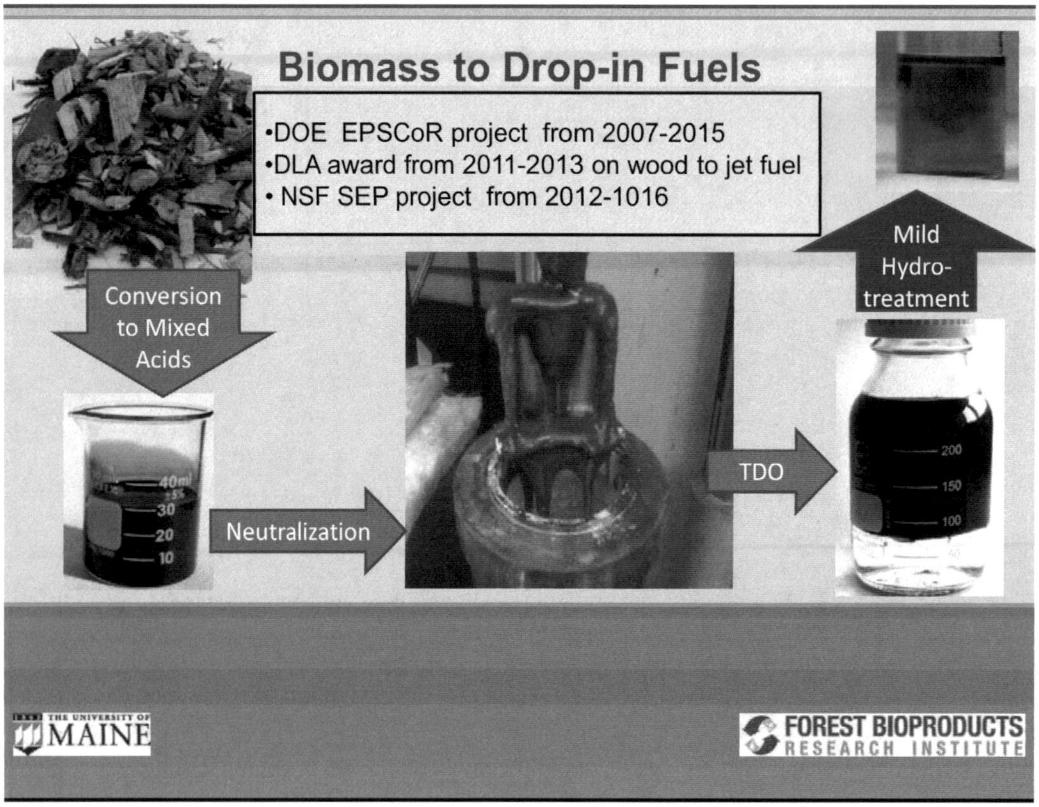

Feedstock Compositions: C6, C5, lignin & Others

	Corn stover	Sugarcane	Wheat straw	Hybrid poplar	Switchgrass
C_6 carbohydrates	41.00	41.21	34.55	46.08	35.71
C_5 carbohydrates	24.00	24.95	22.12	18.58	26.43
Lignin	18.00	23.90	17.28	27.29	19.78
Others	17.00	9.94	26.05	8.04	18.08

	Loblolly pine	Aspen wood
C_6 carbohydrates	54.80	45.80
C_5 carbohydrates	7.30	16.60
Lignin	29.50	21.60
Others	8.40	16.00

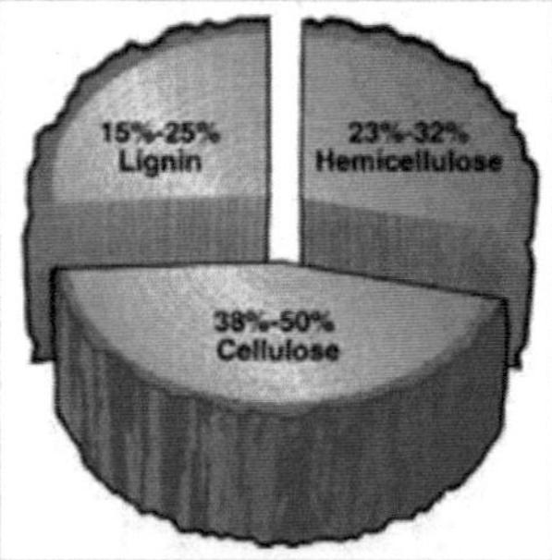

Need to use C6 and C5 Carbohydrates as well as Lignin

THE UNIVERSITY OF MAINE

FOREST BIOPRODUCTS RESEARCH INSTITUTE

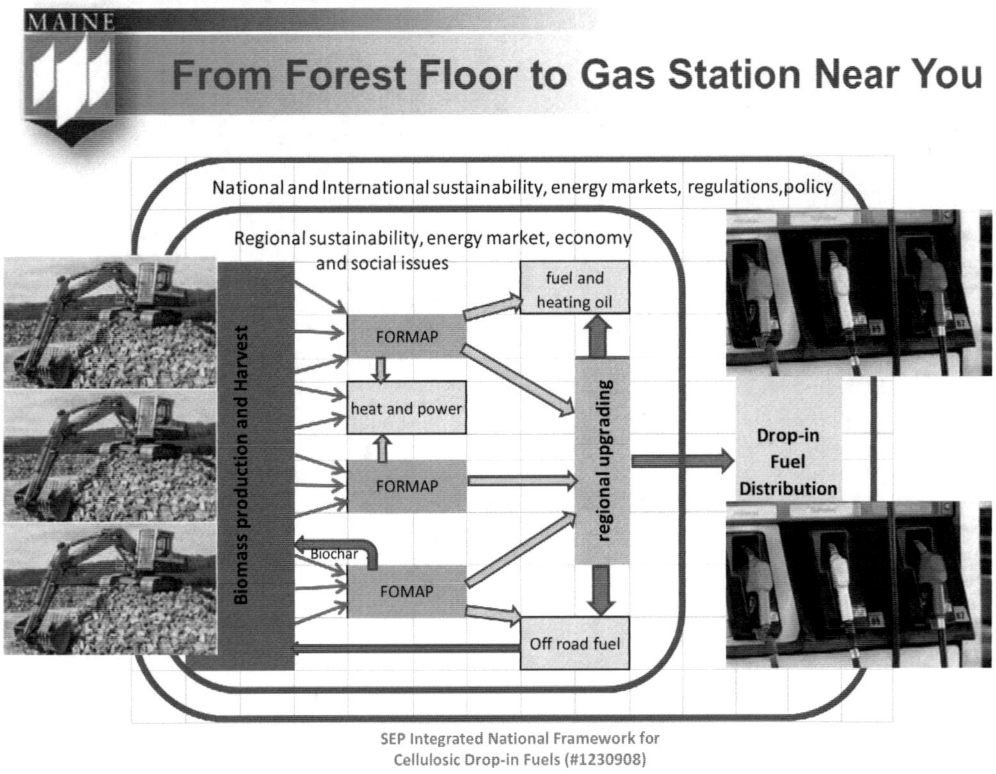
MAINE
From Forest Floor to Gas Station Near You
National and International sustainability, energy markets, regulations, policy
Regional sustainability, energy market, economy and social issues
Biomass production and Harvest
FORMAP
heat and power
FORMAP
Biochar
FOMAP
fuel and heating oil
regional upgrading
Off road fuel
Drop-in Fuel Distribution
SEP Integrated National Framework for Cellulosic Drop-in Fuels (#1230908)

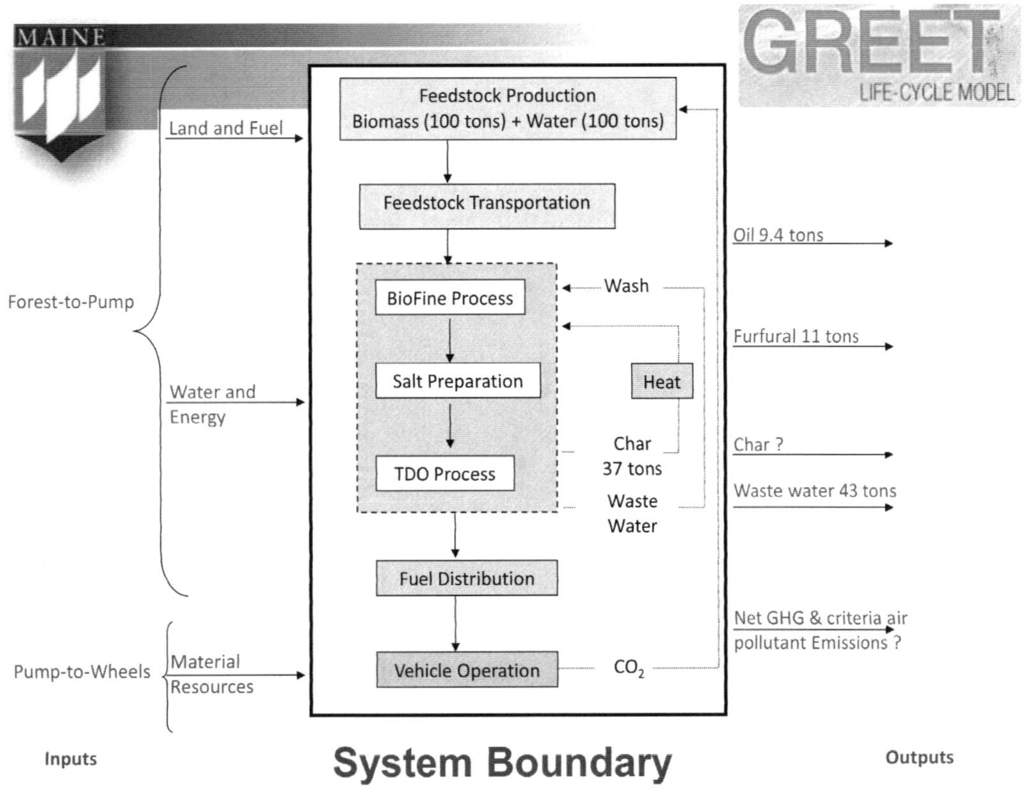
MAINE
GREET LIFE-CYCLE MODEL
Forest-to-Pump
Land and Fuel
Water and Energy
Pump-to-Wheels
Material Resources
Feedstock Production
Biomass (100 tons) + Water (100 tons)
Feedstock Transportation
BioFine Process
Salt Preparation
TDO Process
Wash
Heat
Char 37 tons
Waste Water
Fuel Distribution
Vehicle Operation
CO_2
Oil 9.4 tons
Furfural 11 tons
Char ?
Waste water 43 tons
Net GHG & criteria air pollutant Emissions ?
Inputs
System Boundary
Outputs

Analysis of Biomass Supply Chain

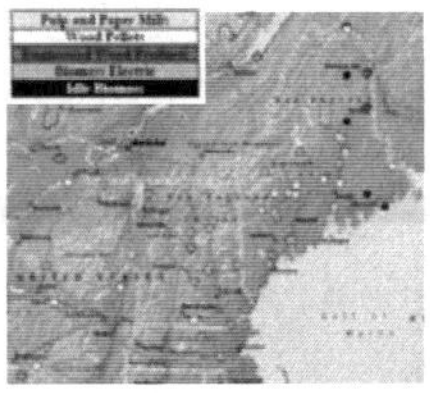

Model the cost and environmental impacts of different biomass harvesting scenarios

Model the cost and environmental impacts of different biorefinery scenarios

Model the cost and environmental impacts of biomass transportation from forest to biorefinery; employ mapping methods to minimize transportation costs

Model the cost and environmental impacts of biofuel transportation from biorefinery to distribution; employ mapping methods to minimize transportation costs

Cellulose to Hydrocarbon Oil

$$C_6H_{10}O_5 \rightarrow 4CH_2 + H_2O + 2CO_2$$

- **Two-step pathway to hydrocarbon oil (Hydrolysis & Dehydration + TDO)**
  - Doesn't require a pure acid intermediate
  - Atmospheric pressure
  - No hydrogen
  - No precious metal or other solid catalysts needed
  - Basic cation (Ca) may be regenerated in pulp mill

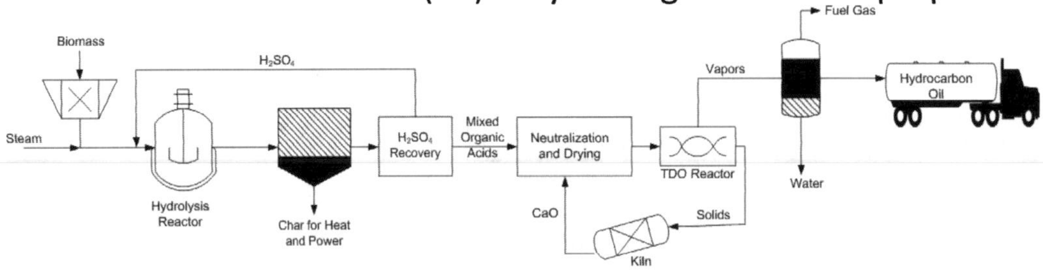

1. Early Discovery:
Thermal Deoxygenation of Levulinic Acid

Energy densification of levulinic acid by thermal deoxygenation, Schwartz et al., Green Chem., 12, 1353-1356 (2010) DOI: 10.1039/c005067a

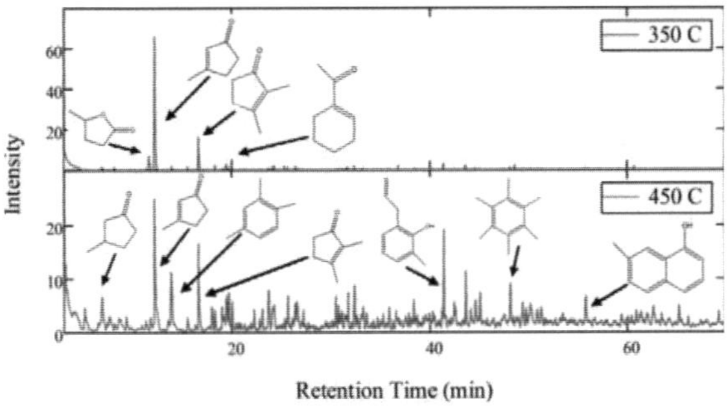

This communication presents a process for thermal deoxygenation (TDO) to upgrade levulinic acid whereby levulinic acid is neutralized with calcium hydroxide, and the resulting salt is heated to between 350 °*C and 450 °C. The* cyclic and aromatic products have low oxygen to carbon ratios, improved energy density, and may be candidates for upgrading to hydrocarbon fuels.

U.S. Patent 8,362,306 (2013) "Energy densification of biomass-derived organic acids"

2. Innovation:
Role of Formic Acid

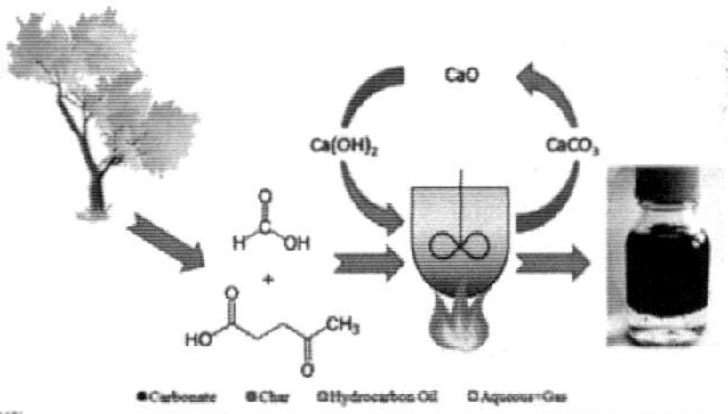

High yields of deoxygenated hydrocarbons are produced by thermal decomposition of formic and levulinic acid mixtures where the optimum feed stoichiometry is consistent with that of cellulose hydrolysis and dehydration. The method promises a high-yield, robust, low-pressure, non-catalytic route for converting biomass hydrolyzates to hydrocarbon mixtures which are similar to petroleum crude oils.

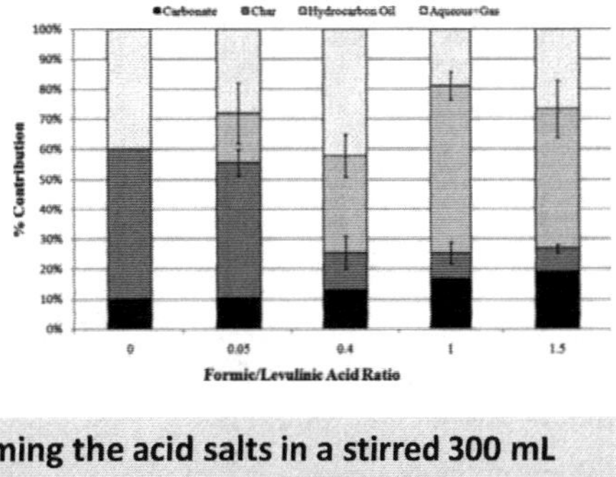

Reactions were conducted by forming the acid salts in a stirred 300 mL Parr reactor along with stainless steel ball bearings to aid in heat transfer and material mixing.

Liquid hydrocarbon fuels from cellulosic feedstocks *via* thermal deoxygenation of levulinic acid and formic acid salt mixtures
Case et al. , *Green Chem.*, 14, 85-89 (2012) DOI: 10.1039/C1GC15914C

3. *Proof of Concept*: Characterization and Combustion of Crude TDO Oil

In this work, TDO oils were characterized for fuel and material properties according to ASTM methods. The oils were found to have low total acid number (<1.4) with oxygen <2%, and a consistent boiling point distribution according to high temperature simulated distillation (ASTM 7169) of between 75° and 585 °C.

Engine tests were conducted at the Oak Ridge National Laboratory Fuels, Engines, and Emissions Research Center (FEERC) on a naturally aspirated, air-cooled Hatz 1D 50z single-cylinder diesel engine with mechanical fuel injection.

Thermal Deoxygenation reactions were performed in a 3L Parr bomb reactor in semibatch operation, then scaled-up to 50L.

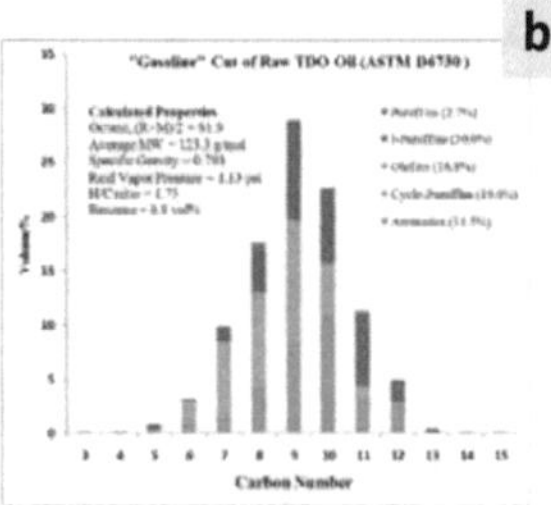

The sub-200°C boiling point cut has 82-84 octane, with specific gravity of 0.8. **RBOB! Eaton et al., Energy Fuels, 27, 5246–5252, (2013)**

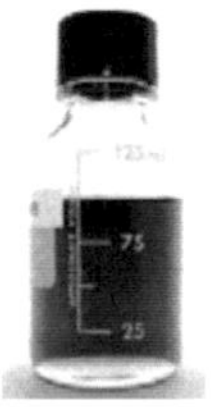

4. *Floor-scale* Synthetic Crude Oil Production, Hydrotreatment, and Fractionation Setup

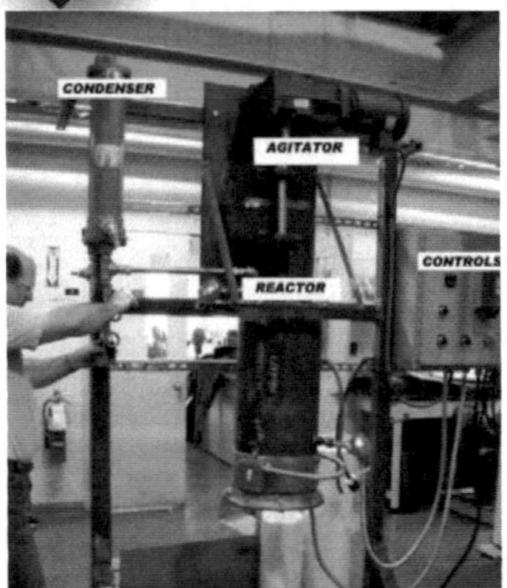

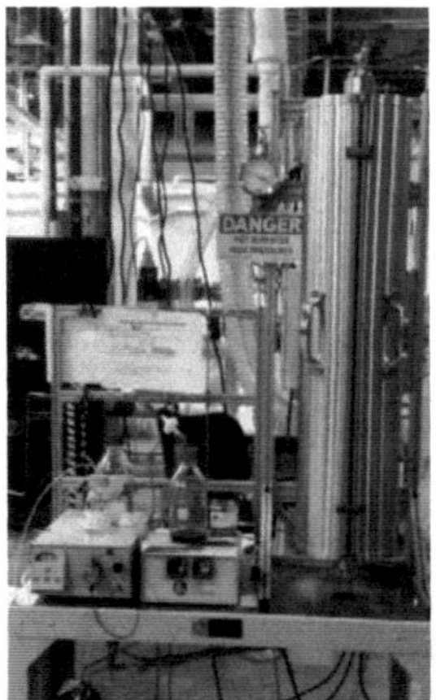

Floor-scale 50 L TDO reactor **ASTM Distillation system**

Hrdrotreatment Reactor

UMaine JP-8 sample tests from AFRL

The AFRL tests confirmed that the jet fuel sample derived from TDO oil is primarily naphthenic, with the density which is "out of spec" at ~0.86 to 0.87 g per mL compared to the maximum spec limit of of 0.84, and the measured heat of combustion at 42.5 MJ/kg which is a bit below the minimum spec limit of 42.8. However, it just means the fuel would have to be blended with conventional jet (or one of the low density alternative jet fuels). The freeze point is found to be less than -70 °C, and low temperature (-20 °C) viscosity is found to be 7.6 cSt which is below the limit of 8 cSt.

Test data on 180 to 250 °C boiling point cut of hydroprocessed TDO Oil

Specification	MIL-DTL-8133H	Measured
Total acid number (mgKOH/g)	★ <0.015	0.002
Aromatics (vol%)	★ <25	11.7
Napthalene (vol%)	<3.0	0.12
Sulfur (mass%)	★ <0.30	<0.100
Flash point (°C)	★ >38	67.5
Density (kg/m^3)	0.775-0.840	0.8745
Freezing point (°C)	★ <-47	<-60
Viscosity (mm^2/s)	★ <8	7.373
Net Heat of Comb (MJ/kg)	>42.8	42.638
Hydrogen Content (wt%)	>13.4	12.88
Smoke point (mm)	★ >25.0	26
Copper Strip Corrosion	★ <No. 1	1a
Existent gum (mg/100mL)	★ <7.0	<1

Environment, Energy, & Economy

- Certified sustainable forest management (with heavy reliance on diesel consumption for harvesting and transportation of biomass) with existing biomass supply infrastructure
- Conversion process, involving recycle of sulfuric acid, water, furfural, and calcium oxide, needs no fossil fuel energy purchase
- Gasoline cut of raw TDO oil is suitable as a low octane blendstock for oxygenate blending (BOB), with diesel cut possibly suitable for off-road or marine applications
- Hydrotreated middle distillate cut yields F76 & HDCD diesel, and Jet A fuel
- Reliance on revenue from furfural and char may be problematic as only the six carbon complex sugars are converted into hydrocarbon liquids

Bringing Research to Marketplace through Rapid Commercialization

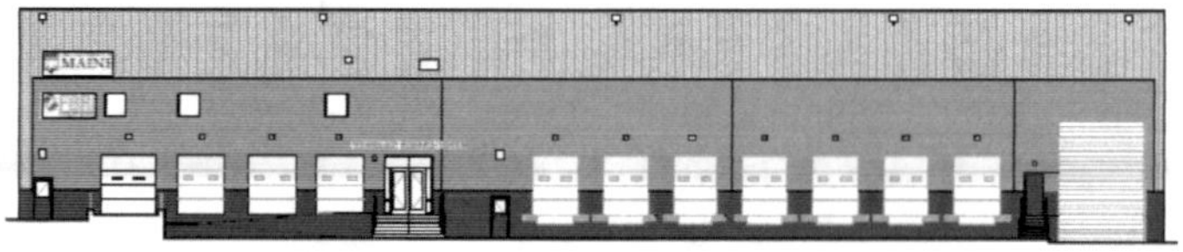

Fuels

Chemicals

Advanced Materials

High Throughput Automation, Data Analysis to Advance Process Development.

Mack, B. C., Bristol-Myers Squibb Company

Originally presented on: 6/12/2014 13:30:00 - 13:55:00

This presentation describes statistical experimental design approaches for high throughput automated experimentation, and the resulting data analysis to guide process development of active pharmaceutical ingredients and intermediates. Laboratory automation drastically decreases the cost and effort associated with routine experimentation during process development. The experimental setup supports development and optimization of multiple unit operations, and includes automated charging, automated sampling, sufficient mixing, and temperature control. To fully realize the value of automated experiments, the experiments must be designed and analyzed in a way that balances speed and rigor, so that experimental findings can be rapidly incorporated into future experimentation. Experiments are designed using various statistically relevant strategies including factorial, screening, and optimal designs. The design helps to strategically allocate the experimental budget for a particular round of automated experiments, which are executed in parallel. Once the data is collected, it is rapidly analyzed by combining data visualization, statistical analysis, and modeling to transform the experimental results into usable process knowledge and guide further rounds of automated experimentation. Once the process design space has been thoroughly investigated, in-depth data analysis and modeling is used to guide experiments and pilot plant batches. A case study where this workflow was applied to an API reaction to guide scale up experimentation and plant condition selection is presented. The case study will discuss the practical implementation of the workflow, the challenges of using small scale models to guide larger scale experiments, and how the results were used to select robust, high yielding operating conditions.

Impact of PAT on Process Development at GSK

American Institute of Chemical Engineering
2014 Process Development Symposium

Gregory Gervasio
June 12, 2014

PAT is used throughout drug development and manufacturing to gain insight and control

Liquid, Solid & Vapor - Included Examples	Raw Materials	Reaction Monitoring	Phase Seperation	Distillation	Crysilization	Drying	Wet Granulation	Milling & Sizing	Blending	Tablet Coating	Tablet Analysis
Mid IR DATR		L		L	L						
NIR	L, S	L, S		L	L, S	S, V	S		S		S
Raman	L, S	L, S		L	L, S	S					S
UV		L		L	L						
Mass Spec		V		V		V					
Particle Vision Measurement					S		S				
Focused Beam Reflectance Measurement					S	S	S				
Laser Diffraction								S			
Other			L							S	S

Reaction Monitoring using Raman Spectroscopy
Two phase water/organic reaction

Raman reference Spectra

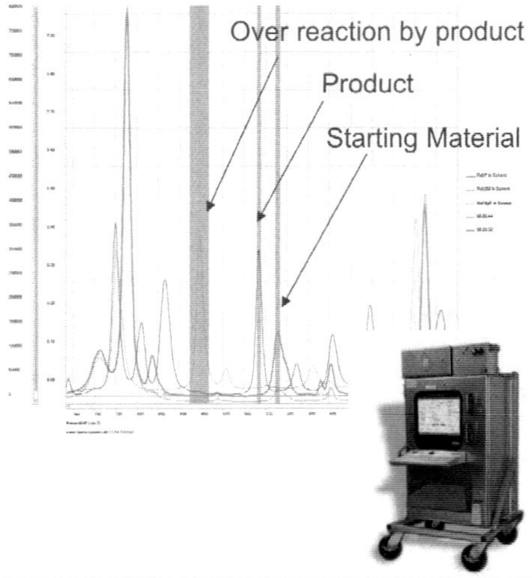

Reaction trend

Reagent addition metered in to controlled reaction

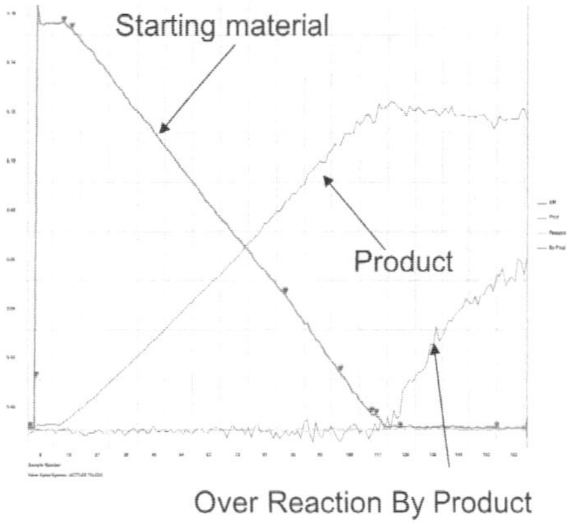

Transfer to Manufacturing

1st Batch inadequate mixing

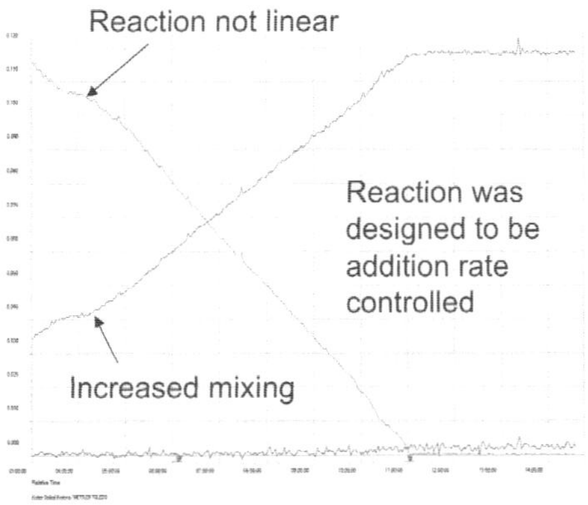

Typical batch

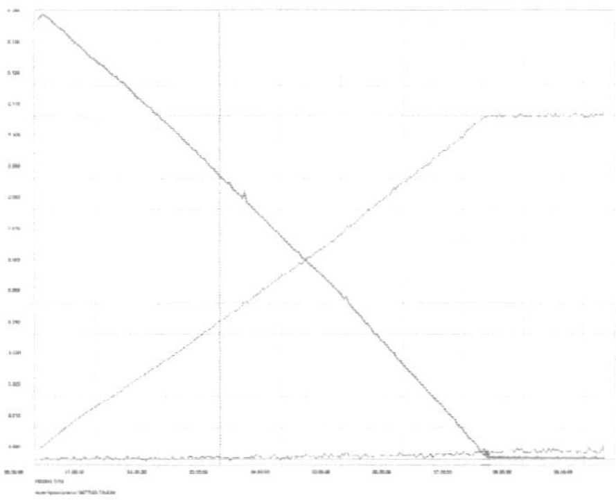

Using Reaction Rate to Predict Reaction Completion

Even using smoothing the small changes in Abs make the rate unusable

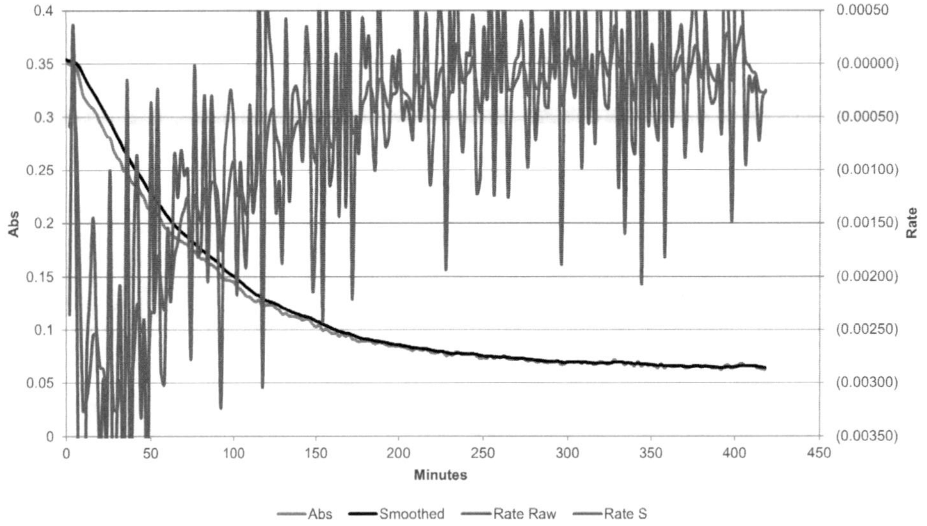

Using Reaction Rate to Predict Reaction Completion

Using Curve fitting on the final data set

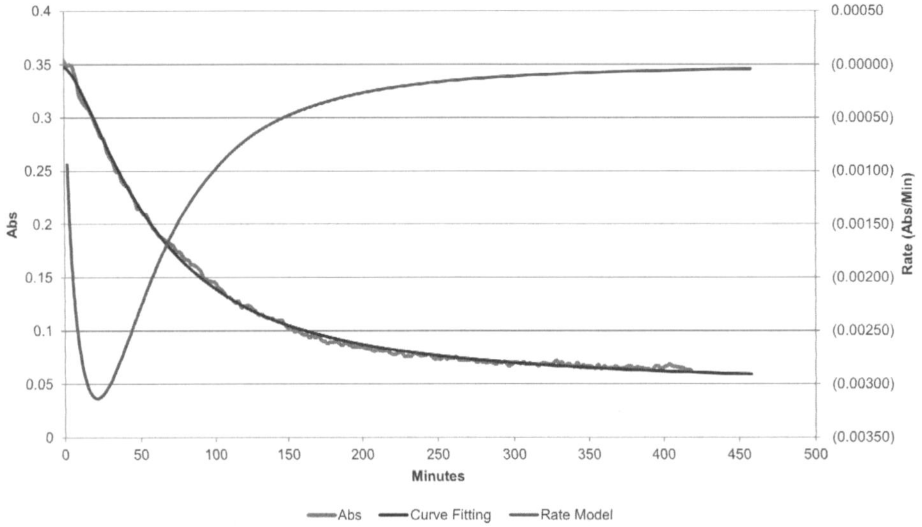

Using Reaction Rate to Predict Reaction Completion

Curve fit recalculated after each point

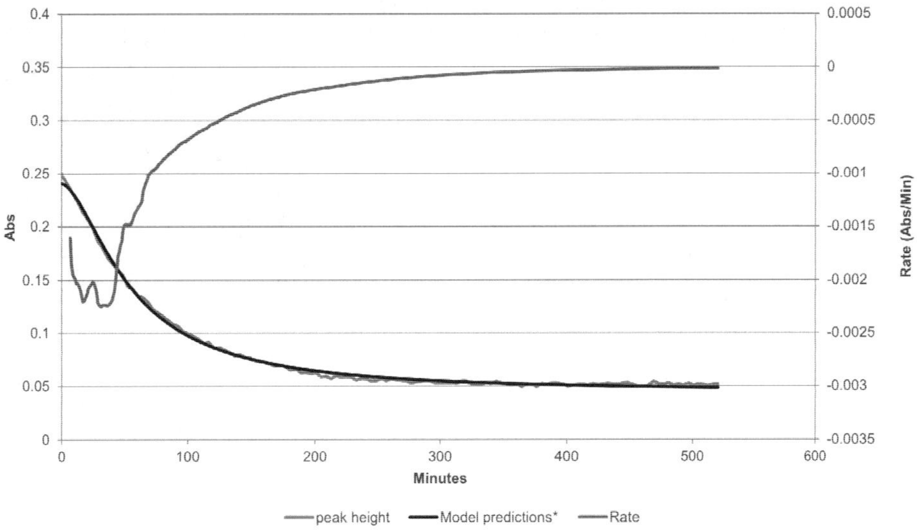

Distillation Monitoring/Control
Direct Measurement via IR

Solvent Swap from Ethanol to Methylcyclohexane

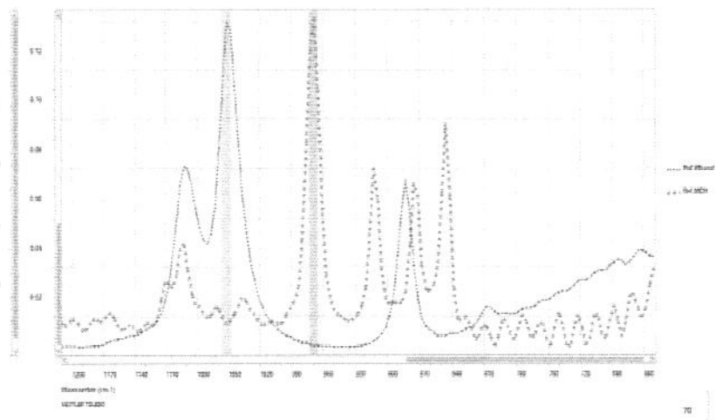

Reference Spectra

Constant volume vacuum distillation
Ethanol replaced with MCH

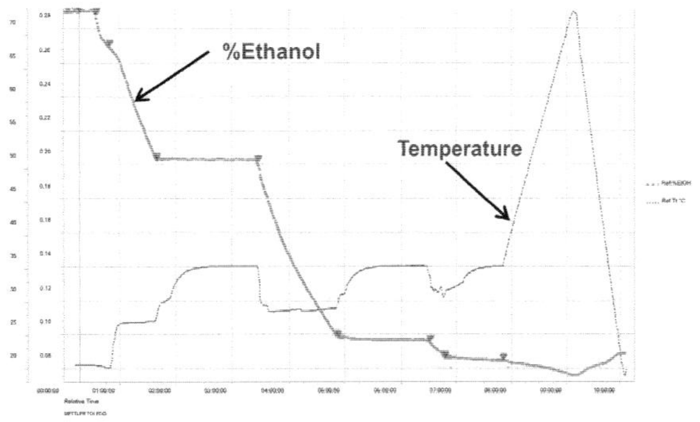

Distillation Monitoring/Control
Condensate - Inferred Measurement Using Vapor Liquid Equilibrium

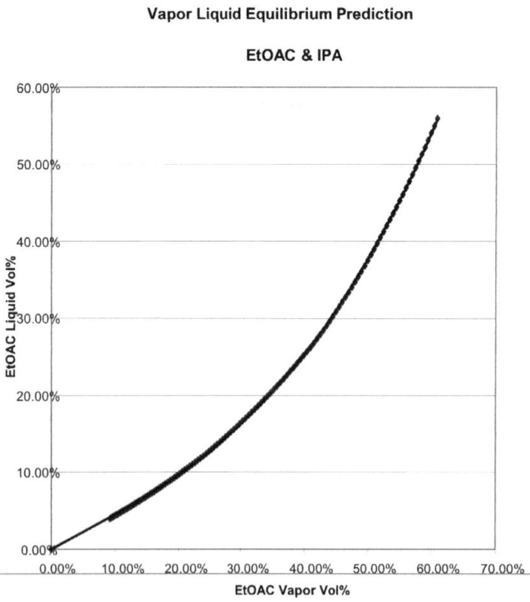

NIR Transmission probe in condensate line

Distillation Monitoring/Control
Condensate - Inferred Measurement Using Vapor Liquid Equilibrium

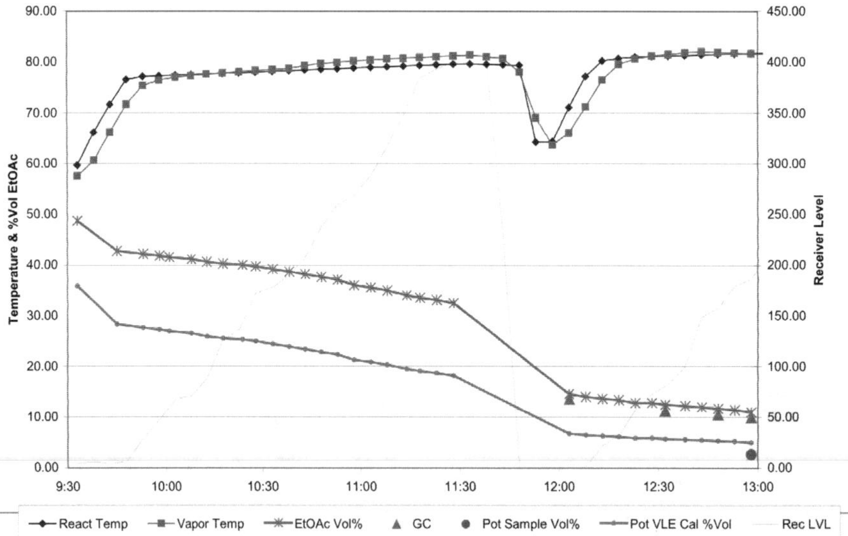

Crystallization Control

For good crystallization control it is important to seed and maintain growth in the metastable zone.

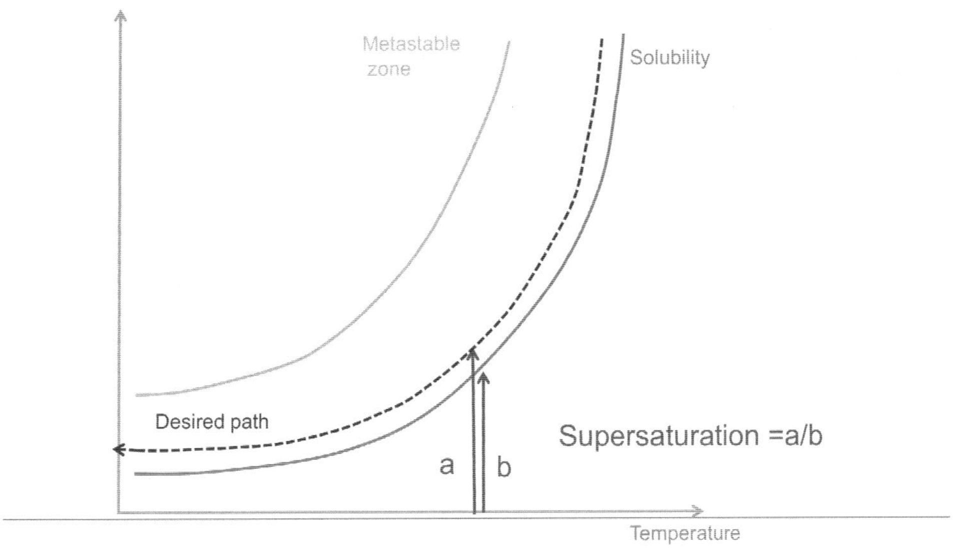

Determining API Solubility & Meta Stable Zone using FBRM

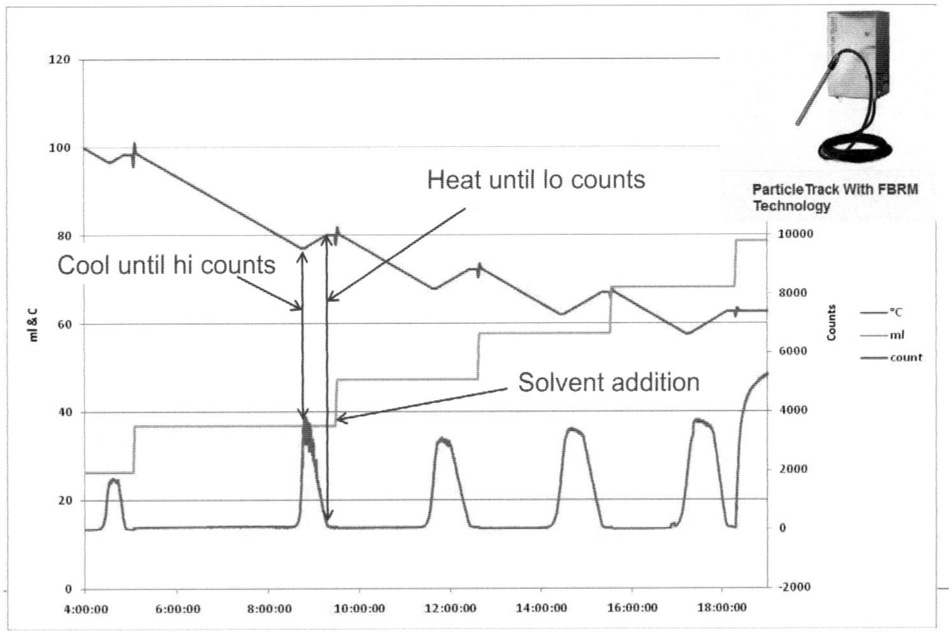

Using IR to understand an Anti Solvent Crystallization

Single experiment was used to determine the IR calibration for both solvents and dissolved API as well as the projected API equilibrium concentration

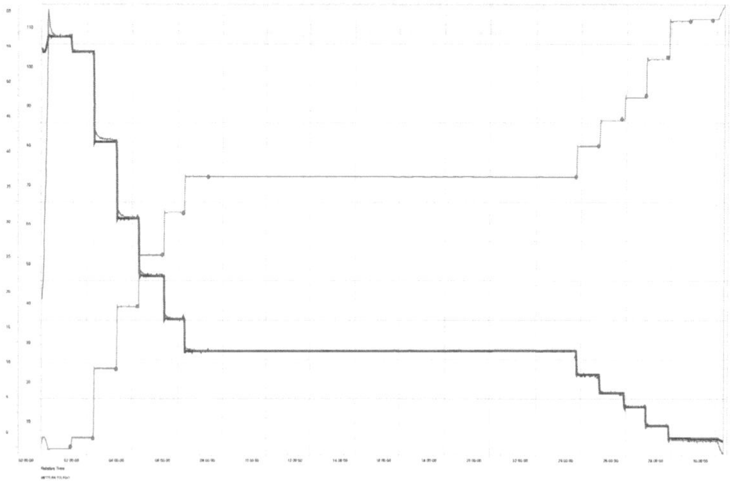

Mid IR with ATR probe

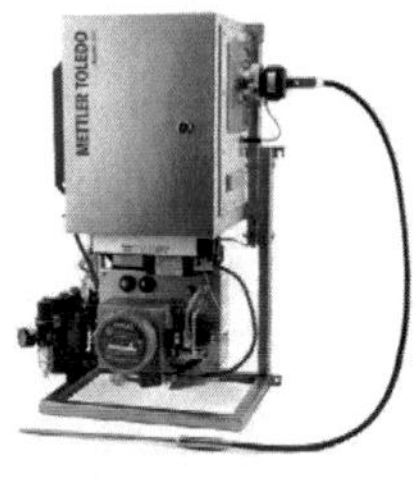

Using IR to understand an Anti Solvent Crystallization

Comparison of three addition rates

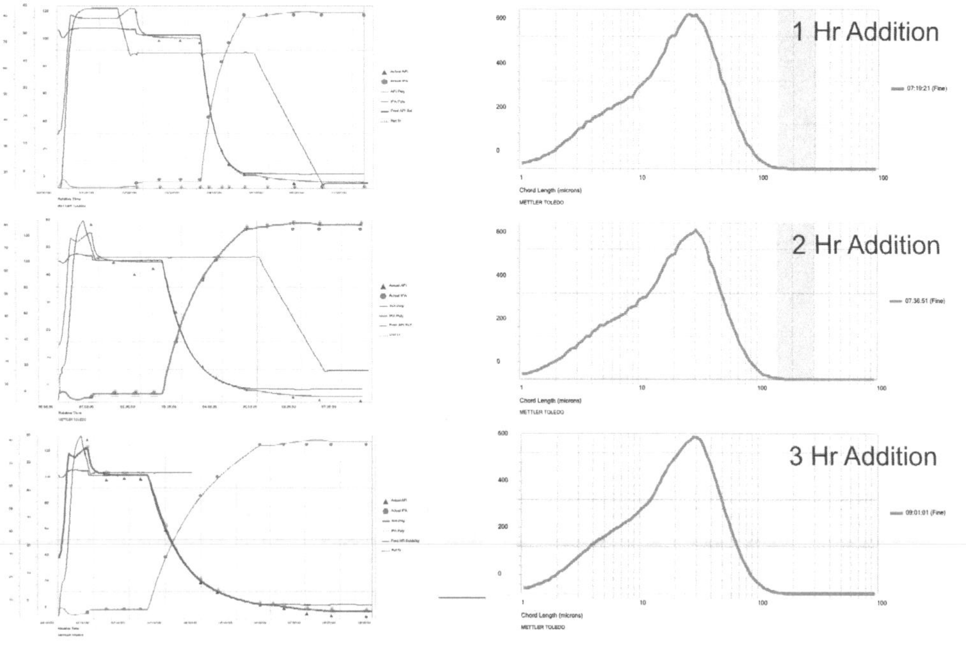

Using FBRM & Raman to Monitor Crystallization & Crystal Form

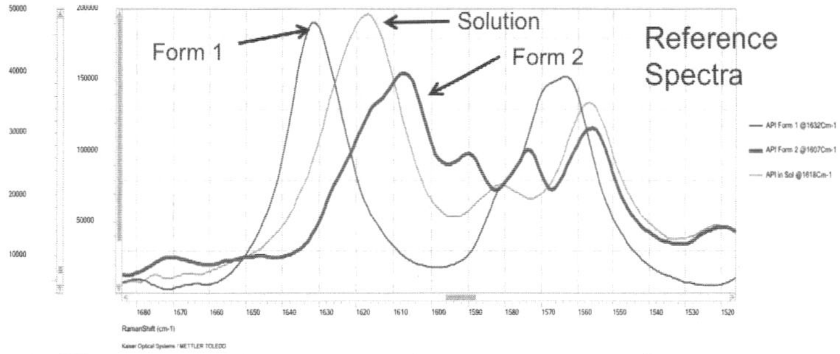

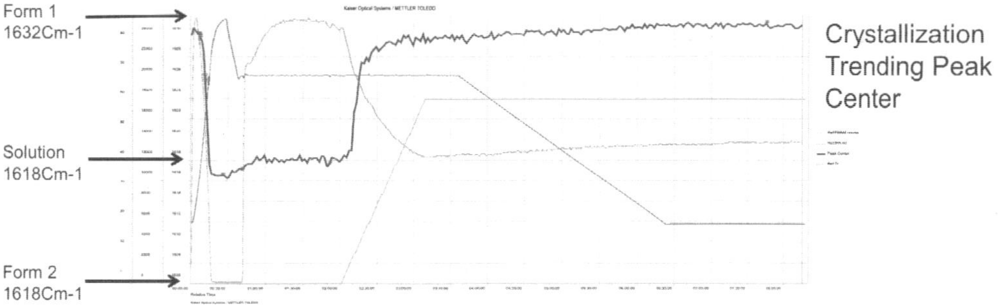

Drying endpoint by Mass Spectroscopy

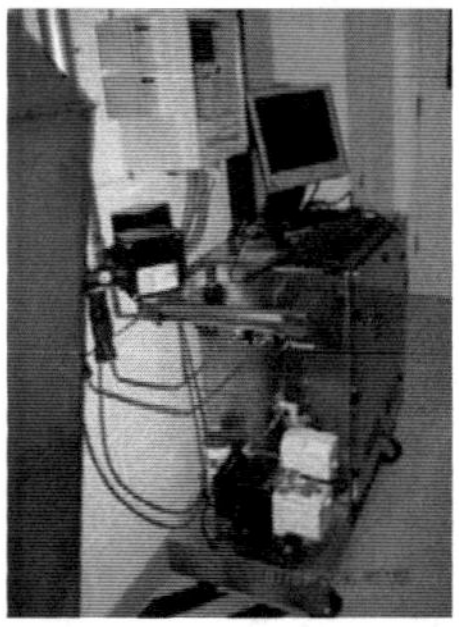

Mass Spectroscopy can monitor multiple solvents and can be multiplexes

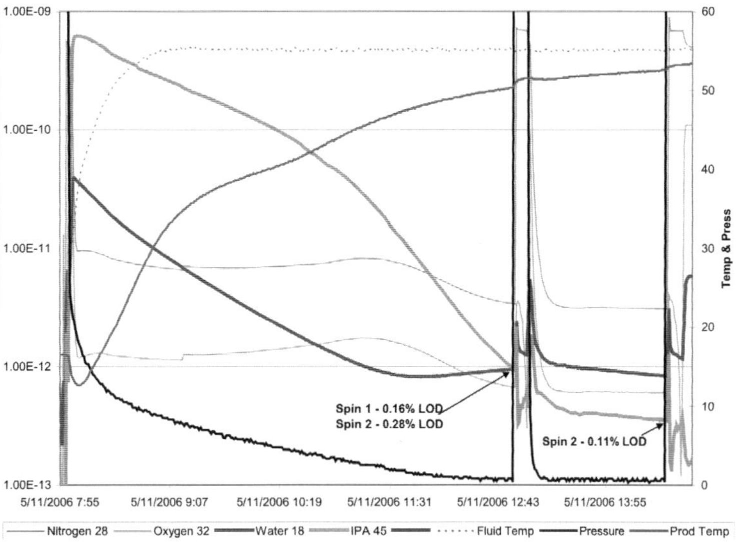

Continuous Granulation

On-Line Analysis of Particle Size and Loss on Drying

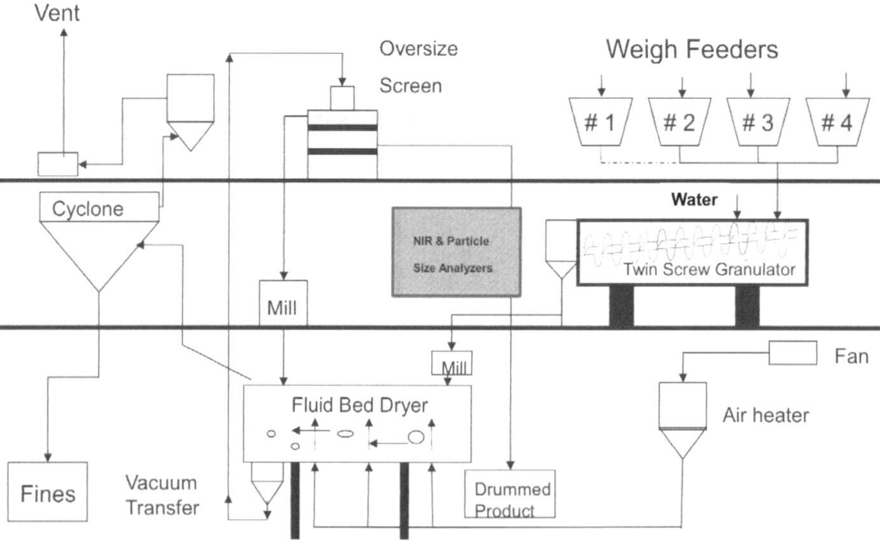

On-Line Laser Diffraction Particle Size Analyzer

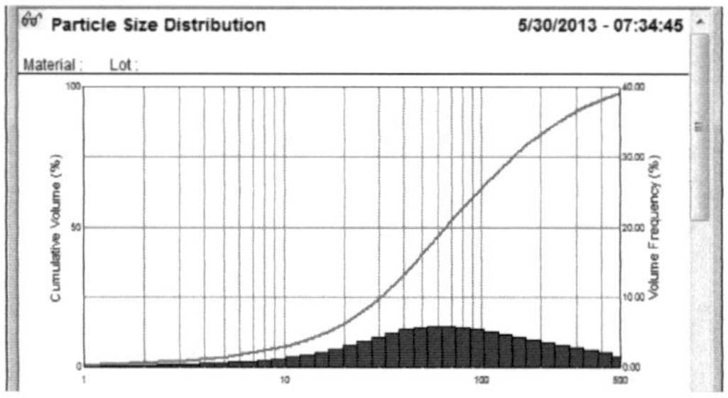

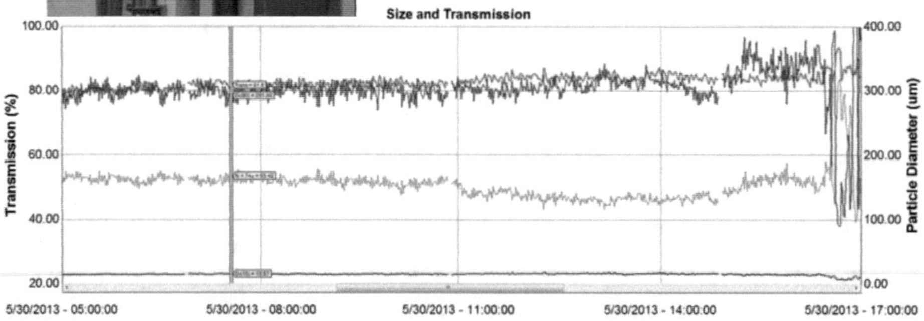

NIR Loss on Drying Analyzer

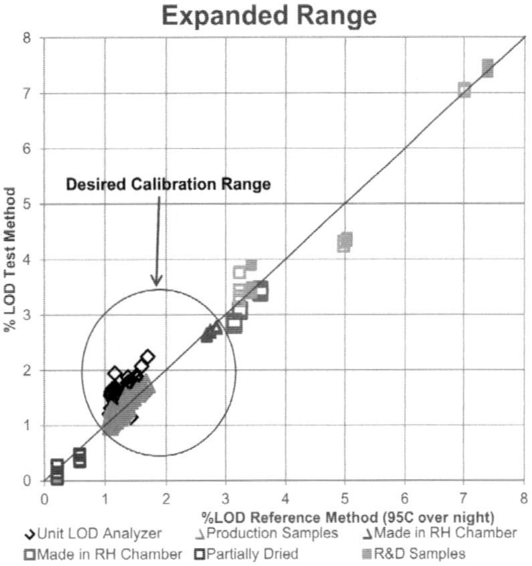

NIR Loss on Drying Analyzer

Sample is collects in cup

Sample is removed by air sweep

Watch out for the unexpected sampling issue

- Using dry instrument air there is a buildup of hard caked product extending from the air sweep nozzle.
- Cause?
  - The cooling effect of the compressed air condensed moisture on the window.
- Solution?
  - Heat air line to raise the temperature above the dew point.
  - Yet to be confirmed

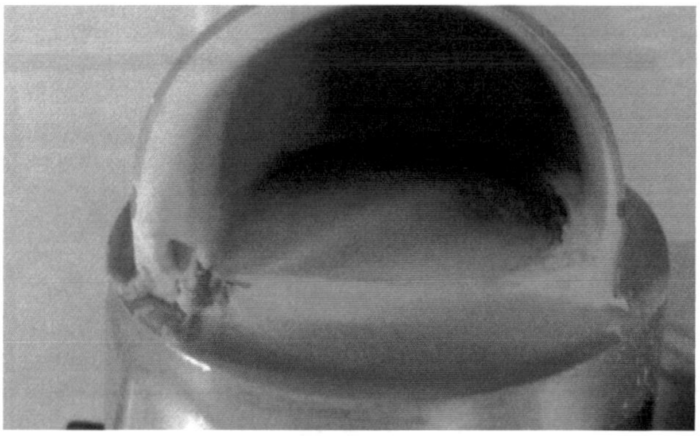

Drug Product Blending Using NIR

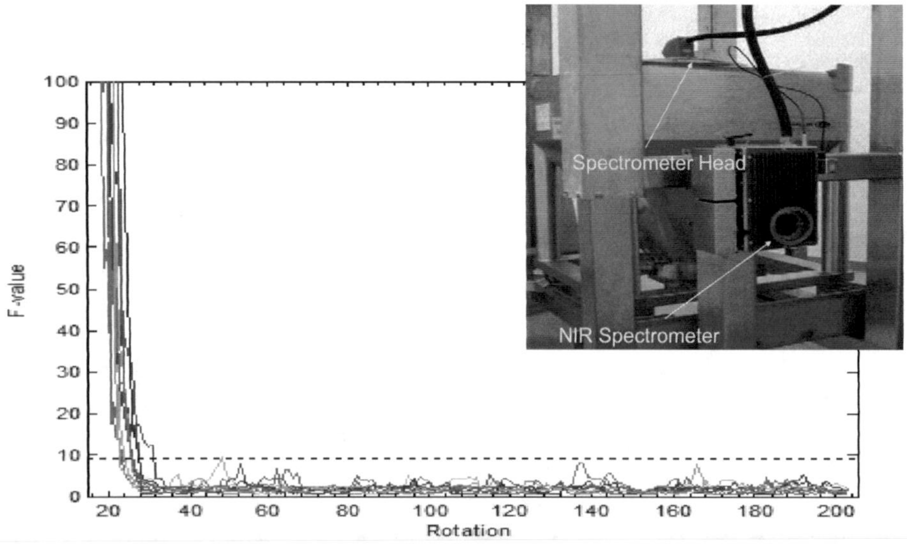

Example Applications – Blending API with Magnesium Stearate

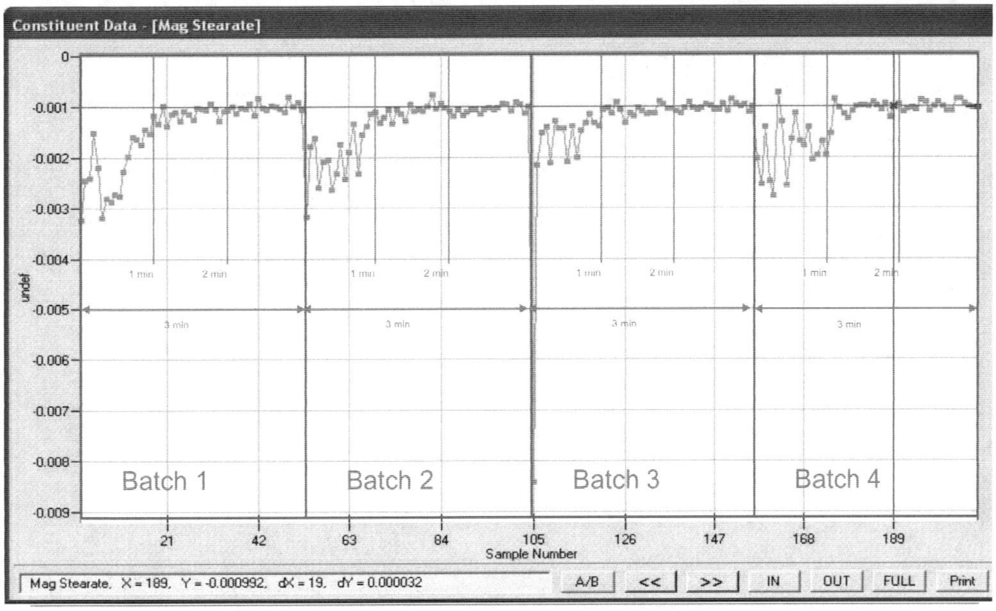

SNV, Discrete point (n=17) second derivative

Near Infrared (NIR) Method of Tablet Analysis

Transmission Measurement

- Bulk sample measurement
- Analyze large portion of sample
- 10 sec to 2 min measurement
- Limited NIR spectral range 7,000-12,500 cm^{-1} (800-1400nm)

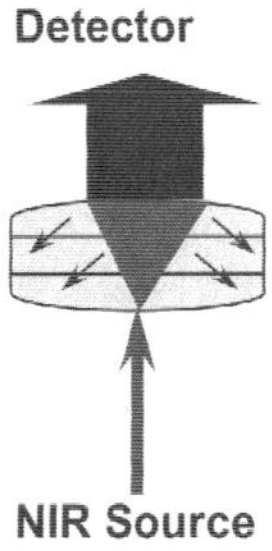

Original Spectra

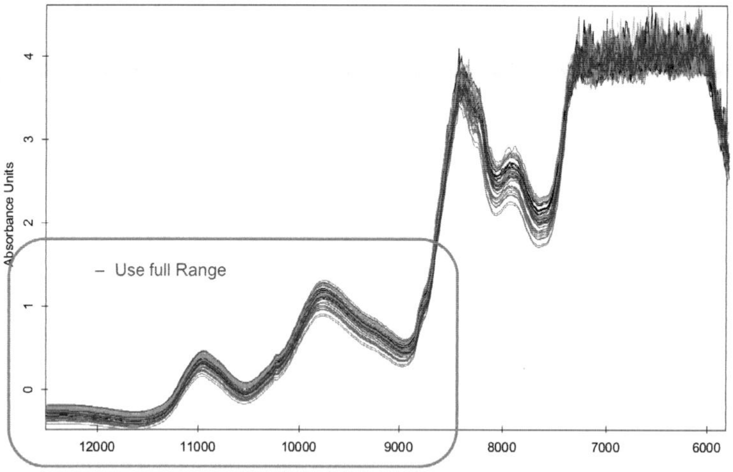

25

API Calibration

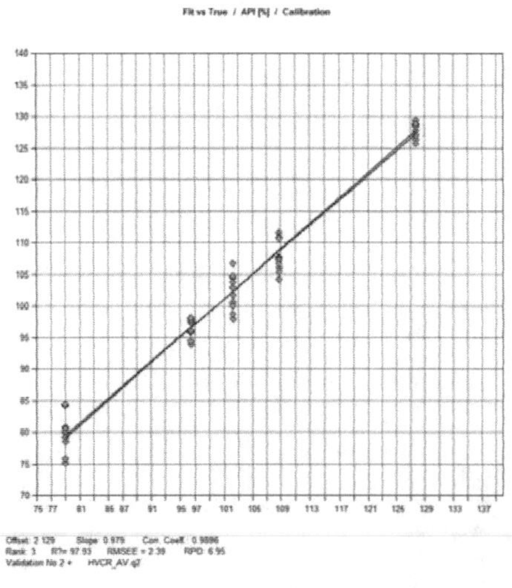

- ❑ Average spectrum of two measurements of each tablet were used to develop calibration.
- ❑ 10 tablets of each strength were used
- ❑ LC concentration values of 79%, 96.5%, 102.25%, 108.75% and 127.75% were used as reference API % values.
- ❑ Partial least square (PLS) was used to develop calibration curve .
- ❑ NIR spectra correlate with API content with R^2 of 97.93, and the accuracy expressed in root mean square error of estimation(RMSEE) is about 2.39% while in root mean square error of leave-one out cross validation (RMSECV) is about 2.63%.

26

On-line Tablet Analysis

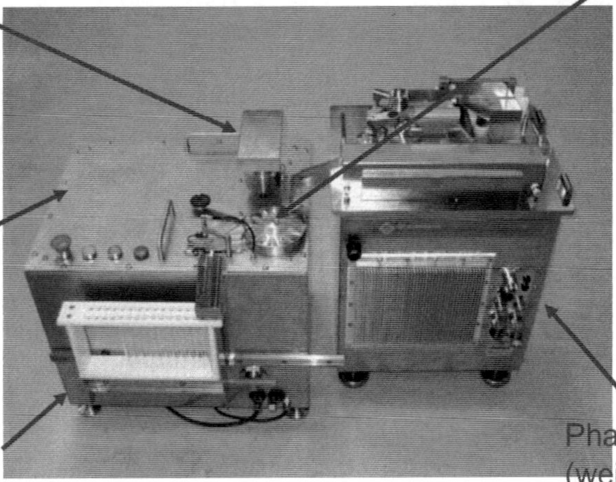

NIR Transmission Content Uniformity analysis

10 position in-feed buffer and accurate tablet handling & positioning

NIR Cube

31 position out-feed tablet storage cassette

Pharmatron 10X-T (weight, thickness, diameter & hardness)

Conclusions

- PAT tools are invaluable for gaining insight to reactions and processes unit operations
- Many PAT technologies exist
- Choose the technology most suited to the application
- Strive for the simplest solution
- A quantitative calibration is not always needed
- Understand the limitations of the technology
- Watch out for sample interface issues
- Get buy-in and support from the end-users regardless of which technique will be used

Acknowledgements & Questions

– Zane Arp - GSK
– Duncan Thompson - GSK
– Vicky He - GSK
– Mettler Toledo

All the Buzz About Process Intensification and Energy Efficiency Through Dow Lens.

Deshpande, K. T., The Dow Chemical Company

Originally presented on: 6/12/2014 14:30:00 - 14:55:00

As the world's population rises and new economies emerge, society requires novel solutions to meet its most basic needs, including energy, water, housing, food, health, and transportation. However, limited natural resources entail their efficient use while simultaneously enhancing the quality of life of current and future generations. In this respect, process intensification is becoming an immensely important tool to achieve overall sustainability of the chemical process industry and the reduction of our energy footprint through controlled plant size.

This talk will focus on two interesting examples pertaining efficient specialty polymer processing as well as efficient synthesis of specialty epoxy through appropriate reactor design. Specifically, single phase solution polymerization renders itself aptly for overall energy reduction by applying process intensification principles. In particular, energy intensive solvent recovery after polymerization results in high energy foot print. Selection of appropriate solvent to lower polymer solubility and form two phases i.e. polymer phase and solvent phase provides an energy efficient alternative. The heavier polymer phase can be separated using a decantation step after polymerization to remove up to 50% of the solvent leading to lower solvent separation costs. Criteria for solvent selection and its influence on process thermodynamics along with impact of processing conditions on important polymer properties will be discussed.

The second example will focus on development of a continuous process for specialty epoxies for lowering the energy footprint. In particular, batch processing of specialty epoxies, though beneficial in controlling temperature of exothermic reactions results in longer processing time and undesirable side products. Manipulating the reactor surface area to volume ratio results in tighter control of reactor temperature and enables safe continuous processing with up to ten-fold reduction in processing time. Development of continuous reactor designs for epoxy processing through modeling and experimental validation will be presented.

Mastering the Art of the Soft Skill

American Institute of Chemical Engineers Symposium

Factor In Human Needs

- Consider needs of people – not just their expertise
- People think, act, perform differently
- They have diverse perspectives, experiences & expectations
- Human interactions will determine success

Establish What to Expect

- Define optimal performance
- Establish expectations
- Team Structure
- Team Processes
- Team Skills
- Shared Understanding
- Support Systems

Team Needs

- Urgency of purpose
- Shared goals
- Sustained commitment to results & to others' success
- Build trust
- Create shared stories
- Awareness of & appreciation for style differences

To Prevent Problems

- Who will lead?
- Roles & responsibilities
- How to communicate
- Necessity for collaboration
- Expectation of feedback
- Ground rules
- Take risks; be creative
- Share ideas

Simple Solutions

- **Good leadership**
- **Manage conflict**
- **Avoid feelings of isolation**
- **Assign tasks & deadlines**
- **Hold regular meetings**
- **Include each person**
- **Ask for input**
- **Delegate**
- **Empower team members**

No Duplication without Permission
JanBara & Associates

Dealing with Cultural Differences

- Not business as usual
- Don't assume you know how to operate
- Ask questions
- Everyone must adjust
- Little information from body language

Dealing with Cultural Differences, con't

- For Example:

- Comfort with directness
- Process toward decisions
- Ease with debate & confrontation
- Informal vs. formal meetings

Questions?

No Duplication without Permission
JanBara & Associates

Tool Kit

No Duplication without Permission
JanBara & Associates

THANK YOU!

If you have further questions, please contact Jane Lowenstein

JanBara & Associates

215 482-1577
owenstein@janbara.com

www.janbara.com

Technology Transfer Documentation the Key to Success.

Vernice, J., Vernice Associates, LLC

Originally presented on: 6/12/2014 15:45:00 - 16:10:00

Successful commercialization of new products is the life blood of business. Many industries institutionalize the preparation of Technology Transfer Documents. This is specifically true for Industries such as the Pharmaceutical, Flavor & Fragrance, and Medical Devices etc. which comply with Good Manufacturing Practice (GMP), and/or International Committee for Harmonization (ICH) regulations. For many businesses, including Small Business, Technology Transfer (TT) Documents are considered "Best Practice". Technology Transfer documents are the common thread in each of these Industries or businesses. In all cases the overall success of the business is tied to the use and quality of the Chemical and Engineering information needed by the stake-holders. It is up to an organization's leaders to create a culture that fosters collaboration. It is these soft skills coupled with organizational leadership that create successful TT, successful products and successful businesses. For companies that depend upon the use of contract development and manufacturing services, the organizational issues and collaboration become even more challenging. This presentation will discuss the TT content and the soft skills needed to produce quality TT Documents. "Best Practice" requires company collaboration to ensure that the stake-holders' needs are met and foster company success.

Communication barriers in long distances.

Yanto, Y., Georgia Institute of Technology

Originally presented on: 6/12/2014 16:15:00 - 16:40:00

At DuPont Crop Protection, we strive to design the next generation of tools to help growers around the world to achieve sustainable farming with improved quantity and quality. We combine global innovation with local collaboration to turn promising ideas into real answers for growers to protect crops such as fruits, vegetables, nuts, corn, soybeans, wheat, and canola. And we focus on ideas that will minimize environmental impact to help preserve our precious resources. In R&D pipeline development, one of the key challenges is moving technology from laboratory research into production facilities. DuPont Crop Protection has focused on building and expanding a network of contract research and manufacturing partners around the world. The alliance network offers high quality science at lower cost, which offers potential for business savings and efficient use manufacturing capacity worldwide. This presentation will highlight examples of our recent work with our external partners in China, India, and Europe. Topics will include: building of product knowledge and scientific competencies with alliance partners, acknowledging and adaption to regulatory challenges in different regions, and analytical support advancement for anti-counterfeit efforts.

AUTHOR INDEX

AIChE
120 Wall Street, FL 23
New York, NY 10005-4020

ISBN 978-1-5108-1759-3

9 781510 817593

CURRAN ASSOCIATES INC.
proceedings
.com